Effects of Flood Control and Other Reservoir Operations on the Water Quality of the Lower Roanoke River, North Carolina

By Ana María García

Prepared in cooperation with the U.S. Army Corps of Engineers

Scientific Investigations Report 2012–5101

U.S. Department of the Interior
U.S. Geological Survey

U.S. Department of the Interior
KEN SALAZAR, Secretary

U.S. Geological Survey
Marcia K. McNutt, Director

U.S. Geological Survey, Reston, Virginia: 2012

For more information on the USGS—the Federal source for science about the Earth, its natural and living resources, natural hazards, and the environment, visit http://www.usgs.gov or call 1–888–ASK–USGS.

For an overview of USGS information products, including maps, imagery, and publications, visit http://www.usgs.gov/pubprod

To order this and other USGS information products, visit http://store.usgs.gov

Suggested citation:

García, A.M., 2012, Effects of flood control and other reservoir operations on the water quality of the lower Roanoke River, North Carolina: U.S. Geological Survey Scientific Investigations Report 2012–5101, 36 p.

Acknowledgments

Numerous individuals and agencies provided support for this project. The support of
Frank Yelverton of the U.S. Army Corps of Engineers, Wilmington District, is acknowledged.
Members of the Kerr 216 Water Quality Group provided valuable insight on water-quality issues
along the Roanoke River. Jim Greenfield and Tim Wool of the U.S. Environmental Protection
Agency, Region 4, provided technical assistance in the use of the hydrodynamic
and water-quality models.

Reviewers of this report included Paul Conrads and Jerad Bales, U.S. Geological Survey.
Their comments are appreciated.

Contents

Figures

Tables

Conversion Factors

Inch/Pound to SI

Multiply	By	To obtain
Length		
inch (in.)	2.54	centimeter (cm)
foot (ft)	0.3048	meter (m)
mile (mi)	1.609	kilometer (km)
Area		
square foot (ft^2)	0.09290	square meter (m^2)
square mile (mi^2)	2.590	square kilometer (km^2)
Volume		
cubic foot (ft^3)	0.02832	cubic meter (m^3)
Flow rate		
cubic foot per second (ft^3/s)	0.02832	cubic meter per second (m^3/s)

SI to Inch/Pound

Multiply	By	To obtain
Length		
centimeter (cm)	0.3937	inch (in.)
meter (m)	3.281	foot (ft)
kilometer (km)	0.6214	mile (mi)
Area		
square meter (m^2)	10.76	square foot (ft^2)
square kilometer (km^2)	0.3861	square mile (mi^2)
Volume		
cubic meter (m^3)	35.31	cubic foot (ft^3)
Flow rate		
cubic meter per second (m^3/s)	35.31	cubic foot per second (ft^3/s)

Temperature in degrees Celsius (°C) may be converted to degrees Fahrenheit (°F) as follows:

°F=(1.8×°C)+32

Temperature in degrees Fahrenheit (°F) may be converted to degrees Celsius (°C) as follows:

°C=(°F-32)/1.8

Vertical coordinate information is referenced to the North American Vertical Datum of 1988 (NAVD 88).

Horizontal coordinate information is referenced to the North American Datum of 1983 (NAD 83).

NOTE: SI units have been used in this report except for measurements that pertain to management of dam release flows. For these Inch/Pound units have been used as requested by the cooperating agency.

Abbreviations and acronyms

BOD	biochemical oxygen demand
DEM	digital elevation model
DO	dissolved oxygen
EFDC	Environmental Fluid Dynamics Code
FERC	Federal Energy Regulatory Commission
LiDAR	light detection and ranging
SOD	sediment oxygen demand
TMDL	total maximum daily load
USACOE	U.S. Army Corps of Engineers
USEPA	U.S. Environmental Protection Agency
USGS	U.S. Geological Survey
WASP	Water Quality Analysis Simulation Program

Effects of Flood Control and Other Reservoir Operations on the Water Quality of the Lower Roanoke River, North Carolina

By Ana María García

Abstract

The Roanoke River is an important natural resource for North Carolina, Virginia, and the Nation. Flood plains of the lower Roanoke River, which extend from Roanoke Rapids Dam to Batchelor Bay near Albemarle Sound, support a large and diverse population of nesting birds, waterfowl, freshwater and anadromous fish, and other wildlife, including threatened and endangered species. The flow regime of the lower Roanoke River is affected by a number of factors, including flood-management operations at the upstream John H. Kerr Dam and Reservoir. A three-dimensional, numerical water-quality model was developed to explore links between upstream flows and downstream water quality, specifically in-stream dissolved-oxygen dynamics. Calibration of the hydrodynamics and dissolved-oxygen concentrations emphasized the effect that flood-plain drainage has on water and oxygen levels, especially at locations more than 40 kilometers away from the Roanoke Rapids Dam. Model hydrodynamics were calibrated at three locations on the lower Roanoke River, yielding coefficients of determination between 0.5 and 0.9. Dissolved-oxygen concentrations were calibrated at the same sites, and coefficients of determination ranged between 0.6 and 0.8. The model has been used to quantify relations among river flow, flood-plain water level, and in-stream dissolved-oxygen concentrations in support of management of operations of the John H. Kerr Dam, which affects overall flows in the lower Roanoke River. Scenarios have been developed to mitigate the negative effects that timing, duration, and extent of flood-plain inundation may have on vegetation, wildlife, and fisheries in the lower Roanoke River corridor. Under specific scenarios, the model predicted that mean dissolved-oxygen concentrations could be increased by 15 percent by flow-release schedules that minimize the drainage of anoxic flood-plain waters. The model provides a tool for water-quality managers that can help identify options that improve water quality and protect the aquatic habitat of the Roanoke River.

Introduction

The lower Roanoke River, which extends from Roanoke Rapids Dam to Albemarle Sound (fig. 1), is an important water resource for North Carolina, Virginia, and the Nation and supports a large and diverse population of nesting birds, waterfowl, freshwater and anadromous fish, and wildlife, including threatened and endangered species. In addition to providing critical habitat for wildlife, the Roanoke River is used for a variety of other purposes, including water supply, hydropower production, wastewater assimilation, and recreation. The flow regime of the Roanoke River is affected by a number of factors, including flood-management operations at the John H. Kerr Dam and Reservoir, and hydropower operations and tidal effects that extend more than half way up the river (Bales and others, 1993). The relations among river flow, flood-plain water levels, and in-stream dissolved-oxygen concentrations are important but poorly understood. Flooding and flood-plain inundation no longer follow a natural seasonal pattern of large floods in the late winter, occasional floods in the fall, and low flows throughout the remainder of the year, but are primarily governed by upstream reservoir releases, which in turn are dependent on flood control schedules and hydropower operations. Recent studies (Bales and Walters, 2003; Wehmeyer and Wagner, 2011) have shown that flood-plain drainage has substantial effects on in-stream dissolved-oxygen levels. The timing, duration, and extent of flood-plain inundation can have either positive or negative effects on vegetation, wildlife, and fisheries in the lower Roanoke River corridor, depending on the inundation characteristics. Because reservoir control operations of the John H. Kerr Dam primarily control overall flows in the Roanoke River, modifications to the reservoir control operations could alter the water quality of both the Roanoke River and the flood plain through changes in flow as well as frequency and duration of flood-plain inundation.

Potential links between upstream flows and downstream water quality in the Roanoke River can be explored using numerical models. A multiagency modeling study of the

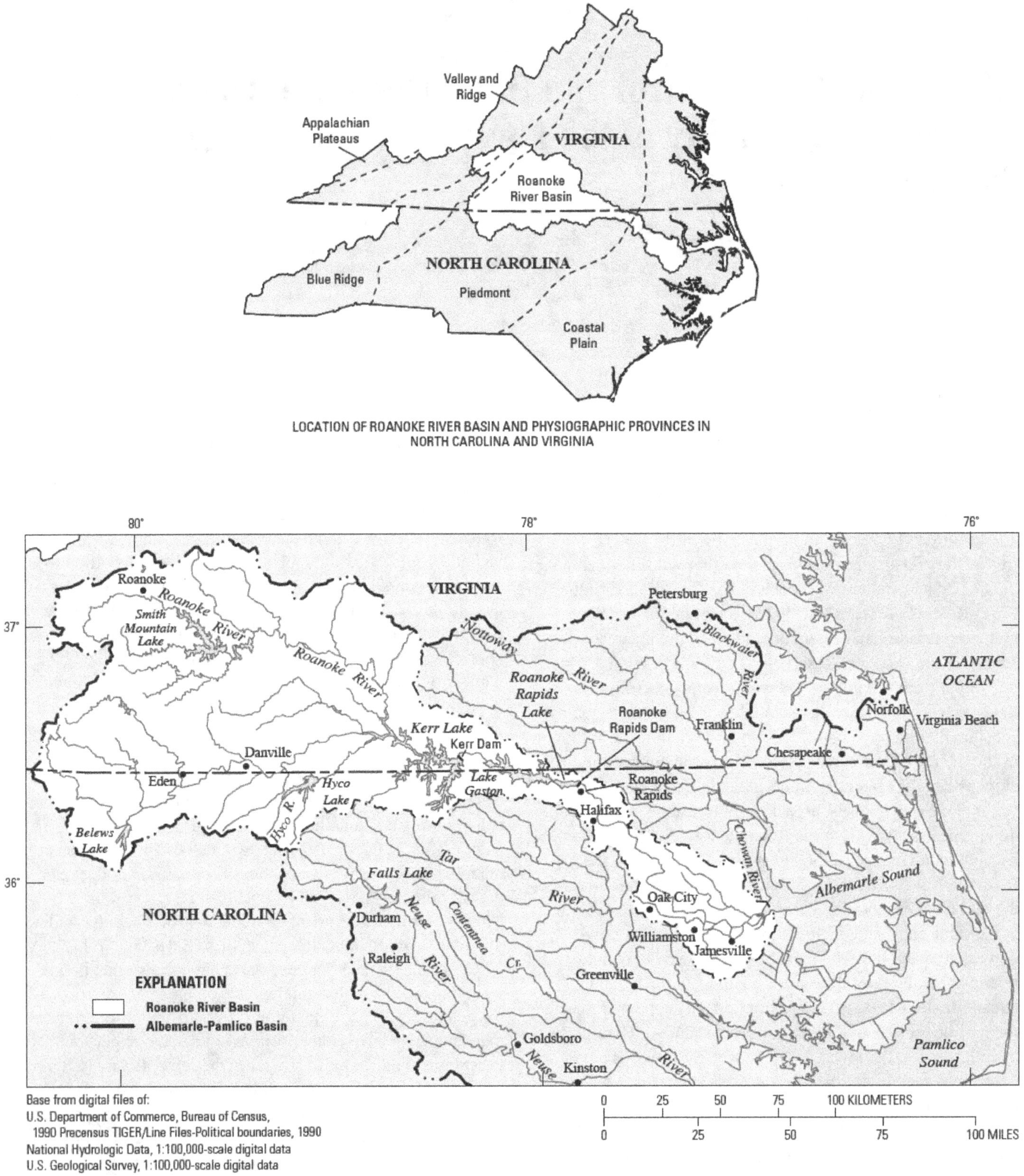

LOCATION OF ROANOKE RIVER BASIN AND PHYSIOGRAPHIC PROVINCES IN
NORTH CAROLINA AND VIRGINIA

Base from digital files of:
U.S. Department of Commerce, Bureau of Census,
 1990 Precensus TIGER/Line Files-Political boundaries, 1990
National Hydrologic Data, 1:100,000-scale digital data
U.S. Geological Survey, 1:100,000-scale digital data

Figure 1.   Roanoke River Basin in North Carolina and Virginia.

Roanoke River was initiated to evaluate the effect of flow-management options on flood-plain inundation and in-stream water quality. Partners in the effort included the U.S. Army Corps of Engineers, North Carolina Division of Water Quality, North Carolina Division of Water Resources, U.S. Fish and Wildlife Service, Dominion Hydropower, and the U.S. Geological Survey (USGS). Although previous studies had applied one-dimensional models to the lower Roanoke (Wehmeyer and Wagner, 2011), a specific simulation of lateral (in and out of the flood plain) flows was needed to simulate the exchange of water and oxygen between the main channel and anoxic swamp waters, a process which has been shown to be an important determinant of main channel oxygen levels (Bales and Walters, 2003). Therefore, two multidimensional numerical models were developed to establish the cause and effect between reservoir operations and in-stream dissolved-oxygen levels and to study management scenarios that maintain water quality and support aquatic life.

This study supports one of six themes outlined in the science strategy of the USGS "to inform the public and decision makers about … forecasts of likely outcomes for water availability, water quality, and aquatic ecosystem health caused by changes in land use and land cover, natural and engineered infrastructure, water use, and climate" (U.S. Geological Survey, 2007a). The study also provides support to Federal agencies in meeting their resource management objectives.

Purpose and Scope

This report documents the development and application of a hydrodynamic model and a water-quality model to establish links between operations for flood control and hydropower, and downstream water quality. The geographic area of interest is the extent of the Roanoke River that is affected by management of water levels at the John H. Kerr Dam and Reservoir (Kerr Dam) and by hydropower genera-tion. The river reach extends from the USGS streamgage near Roanoke Rapids (site 1, fig. 2) to the USGS streamgage near NC 45 (site 9, fig. 2).

Alternative flood control scenarios were considered by the U.S. Army Corps of Engineers (USACOE) to modify operations during warm-weather months. The models were configured and calibrated to data collected between June and September 2006 to address whether management changes can substantially effect water quality along the main stem of the lower Roanoke River.

Study Area

The study area is the lower Roanoke River, which extends from the USGS streamgage near Roanoke Rapids to the USGS streamgage near NC 45 near Batchelor Bay (fig. 2), a distance of about 220 river kilometers (km). The lower Roanoke River Basin is about 3,620 square kilometers (km^2) and is located in the Coastal Plain Physiographic Province (fig. 1) of northeastern North Carolina. The lower Roanoke River flood plain ranges in width from 5 to 10 km and covers 90 percent of the river length. Flows in the Roanoke River are affected by several human-induced and natural factors, including reservoir releases at both the Kerr Dam and at the Roanoke Rapids Dam, hydropower peaking and sustained releases, flood-plain storage and subsequent drainage, evapotranspiration from the flood plain, and tidal backwater in Albemarle Sound.

The study area includes some of the largest and least disturbed broad expanses of wetland forest on the eastern seaboard of the United States. These forests can be classified on the basis of geography and the duration of inundation or hydroperiod (period of time during which the forest or wetland is covered by water). Forested peatlands are found at the mouth of the Roanoke River and the Cashie River, an area that is at or very near sea level. Swamp forests, typical of the Southeastern United States, can be found near Williamston, N.C., and are characterized by having long hydroperiods. Near Oak City, N.C., bottomland forests are either wet bottomland hardwoods, which typically are flooded every year for varying periods of time, or mesic bottomland forests, which are flooded infrequently (Townsend, 2000).

Hydroperiods no longer follow a natural pattern for the Roanoke River, but are dependent on flow releases at Kerr Dam and Roanoke Rapids Dam. Kerr Dam operations, in turn, are dependent on a seasonally varying guide curve that controls water levels at the reservoir. The guide curve sets targeted water-surface levels at the reservoir for specific periods of time. The targeted levels are affected by the natural inflow to the reservoir from tributaries, precipitation, and evaporation. Lake levels are also managed to minimize downstream flooding and allow hydropower production. Under existing operations that have been employed to manage lake levels in the past several decades, during flood conditions, the plan requires a discharge of 20,000 cubic feet per second (ft^3/s) (566 cubic meters per second (m^3/s)) for reservoir levels between 300 and 312 feet (91.4 and 95.1 meters) above NAVD 88, which leads to prolonged downstream flood-plain inundation during warm-weather months. When releases drop to minimum levels, subsequent flood-plain drainage can cause sudden drops in dissolved-oxygen concentrations as anoxic waters enter the main channel, causing fish kills.

Data Used for Model Development and Application

Although several streamgaging stations on the Roanoke River have been continuously monitored for water level and dissolved oxygen since 1998, additional sites were monitored during 2006–2007 to provide the data necessary to properly calibrate both a hydrodynamic and a water-quality model. Data collected by the USGS during this period were used (1) for boundary conditions of the models, (2) as calibration targets, and (3) to parameterize physical characteristics of the system.

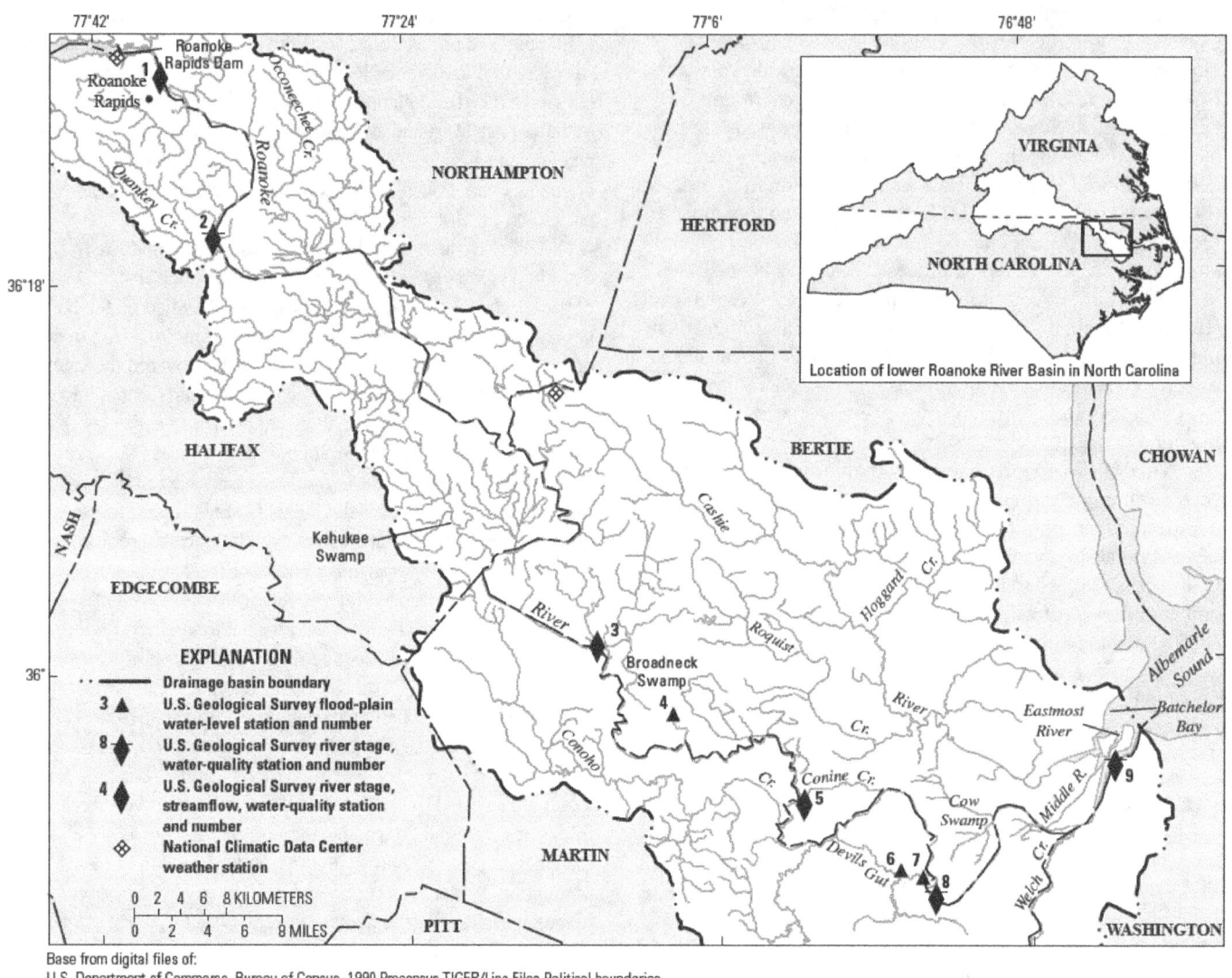

Base from digital files of:
U.S. Department of Commerce, Bureau of Census, 1990 Precensus TIGER/Line Files-Political boundaries,
1990 National Hydrologic Data, 1:100,000-scale digital data
U.S. Geological Survey, 1:100,000-scale digital data

Figure 2. Locations of the U.S. Geological Survey and National Climatic Data Center data-collection sites in the lower Roanoke River Basin, North Carolina.

Boundary Conditions

The hydrodynamic model requires upstream and downstream boundary conditions for the variables simulated. For this study, continuous streamflow data collected between June and September 2006 at the USGS station at Roanoke River at Roanoke Rapids were used as upstream boundary time series. Measurements of water temperature made at Roanoke River near Halifax were used to define the upstream boundary condition for temperature. Continuous water-level and temperature measurements made at the USGS station at Roanoke River at NC 45 constituted the downstream boundary conditions.

For the water-quality model, time series of dissolved-oxygen concentrations measured at USGS streamgages located at the Roanoke River near Halifax and Roanoke River at NC 45 were used as upstream and downstream boundary conditions, respectively. In addition, biochemical oxygen demand (BOD) data gathered at the USGS Roanoke River at Roanoke Rapids streamgage were used to quantify BOD loading of the inflow to the lower Roanoke River.

Table 1. Streamflow, water-quality, and flood-plain level monitoring stations in the Roanoke River Basin, North Carolina, that were used in the study.

[NAD 83, North American Datum of 1983; km, kilometer; km², square kilometer; River km, the distance from the mouth of the Roanoke River to the station; Q, streamflow; BOD, biochemical oxygen demand; T, temperature; DO, dissolved oxygen; WL, water level; EFDC, Environmental Fluid Dynamics Model; WASP, Water Quality Assessment Program; –, not applicable]

Map no. (fig. 1)	Station name	Station number	Latitude (NAD 83)	Longitude (NAD 83)	Station type	River km	Drainage area (km²)	Data used in model	Data purpose	Model	Period of record
1	Roanoke River at Roanoke Rapids	02080500	36 27'38"	77 38'03"	River	208	21,816	Q	Boundary	EFDC	Dec. 1911–present
								BOD	Boundary	WASP	June 2006–Nov. 2007
2	Roanoke River at Halifax, NC	0208062765	36 27'38"	77 38'03"	River	187	22,001	T	Boundary	EFDC	Mar. 1998–present
								DO	Boundary	WASP	Mar. 1998–present
3	Roanoke River near Oak City	02081022	, 36 00'51"	77 12'54"	River	106	22,927	Q	Calibration	EFDC	June 2006–Nov. 2007
								WL	Calibration	EFDC	June 2006–Nov. 2008
								DO	Calibration	WASP	Mar. 1998–present
4	Broadneck Transect 2	355722077082801	35 57'21.73"	77 17'25.68"	Flood plain	–	–	WL	Calibration	EFDC	June 2006–Nov. 2007
5	Roanoke River at Williamston	02081054	35 51'41"	77 02'19"	River	58	23,595	Q	Calbration	EFDC	June 2006–Nov. 2007
								WL	Calibration	EFDC	Dec. 1911–present
								DO	Calibration	WASP	Mar. 1998–present
6	Devils Gut Transect 2	355024076562301	35 50'25"	76 56'20"	Flood plain	–	–	BOD	Parameter	WASP	June 2006–Nov. 2007
7	Devils Gut near mouth near Jamesville	354944076541200	35 49'40"	76 54'12"	Flood plain	–	–	DO	Calibration	WASP	June 2006–Nov. 2007
8	Roanoke River at Jamesville	0201094	35 48'49"	76 53'37"	River	31	24,056	WL	Calibration	EFDC	Dec. 1911–present
								DO	Calibration	WASP	Mar. 1998–present
9	Roanoke River at NC 45 Bridge	02081141150	35 48'49"	76 53'37"	River	2.6	25,143	WL	Boundary	EFDC	Mar. 1998–present
								T	Boundary	EFDC	Dec. 1911–present
								DO	Boundary	WASP	Mar. 1998–present

Calibration Data

Although continuous, long-term water-level data were available at several in-stream sites on the lower Roanoke River, hydrodynamic model calibration was limited to the period of time when streamflow data were available (table 1). Continuous measurements of streamflow and water level made between June and September 2006 were used for model calibration. Because the calibration period includes small fluctuations in temperatures (as compared to a multiseason simulation), temperature data were not used for calibration. Continuous dissolved-oxygen concentrations at the in-stream sites and discrete water-quality samples analyzed for dissolved-oxygen concentrations at one flood-plain location were used to calibrate the water-quality model.

Parameterization of Physical Characteristics

During the June 2006 and September 2007 data-collection efforts, discrete water-quality samples were collected from locations in the Roanoke River flood plain. Some of these data (table 1) were used in the water-quality model to parameterize the BOD of the water column in the flood plain.

Point sources may not be part of the natural physical system, but several facilities that are permitted to discharge into the Roanoke River were represented in both the hydrodynamic and water-quality models. Daily flows, dissolved-oxygen concentrations, and BOD (both 5-day and ultimate) data for the major permitted dischargers were provided by the North Carolina Department of Environmental Quality and used in the construction of the models (table 2).

Table 2.   List of facilities permitted to discharge into the Roanoke River and the amount of permitted flow.

[WWTP, wastewater-treatment plant; m³/s, cubic meters per second]

Permit	Facility	Permitted flow (m³/s)
NC0024201	Roanoke Rapids Sanitary District WWTP	0.365
NC0025721	Weldon Town WWTP	0.053
NC0027626	North Carolina Department of Correction, Caledonia WWTP	0.022
NC0027642	Odom Correctional Institute WWTP	0.005
NC0023116	Lewiston/Woodville Town WWTP	0.007
NC0044776	Town of Hamilton WWTP	0.004
NC0020044	Town of Williamston WWTP	0.074
NC0035858	Town of Jamesville WWTP	0.007
NC0000680	Domtar Paper Company LLC	3.615
NC0020028	Town of Plymouth WWTP	0.035

Description of the Modeling System

Two multidimensional numerical models were applied to the lower Roanoke River to simulate dissolved-oxygen dynamics. Descriptions of the models used in the study are presented in the following section.

Environmental Fluid Dynamics Code (EFDC)

The Environmental Fluid Dynamic Code (EFDC) is a three-dimensional hydrodynamic and transport model supported by the Watershed and Water Quality Modeling Technical Support Center of the U.S. Environmental Protection Agency (USEPA; Hamrick, 1992). The code uses a finite difference scheme to solve the equations of motion and transport for turbulent kinetic energy and temperature in vertical and horizontal coordinate systems, which may be Cartesian or curvilinear-orthogonal. Model stability and convergence can be achieved under highly dynamic boundary conditions, such as variable reservoir release schedules, down-stream tidal effects, and climatic extremes, such as hurricanes. The code used in this study did not include the option of an adaptive timestep; therefore, the size of the timesteps was limited by numerical stability during rapid changes in boundary conditions, leading to very long computational times. For the lower Roanoke River, the model required 1 hour of computational time per month of simulation. The model also has algorithms to simulate wetting and drying within the model domain (Ji and others, 2001), which were configured to simulate Roanoke River flood-plain processes.

The EFDC model has been successfully applied to other riverine systems, including the Neuse River (Wool and others, 2003) in North Carolina and Suwannee River estuary in Florida (Bales and others, 2006). Other applications include simulation of wind- and thermal-driven circulation in Lake Okeechobee (Jin and others, 2000; Jin and others, 2002) and simulation of circulation and salt transport in the Indian River Lagoon in Florida (Moustafa and Hamrick, 1994; Suscy and Morris, 1998).

Water Quality Analysis Simulation Program (WASP)

The water-quality model that was applied to the lower Roanoke River study area is the USEPA Water Quality Analysis Simulation Program (WASP) model, version 7.41 (Ambrose and others, 1993; Wool and others, 2003). The WASP model is widely used by the USEPA, State agencies, and contractors to determine total maximum daily loads (TMDLs), for point-source permitting, and in other types of water-quality investigations. The version of WASP used for this project was version 7.41, which is an enhanced Windows version of the model with features that include a pre-processor and a graphical post-processor. Version 7.41 is also fully coupled with hydrodynamic data from EFDC, allowing for the simulation of mass transport in three dimensions as configured in EFDC.

The model includes five kinetic submodels: eutrophication, toxicants, mercury, thermal, and fecal coliform. The eutrophication submodel, which was used in this application, can be configured at various levels of complexity to simulate some or all of the following dissolved-oxygen processes: (1) Streeter-Phelps, (2) modified Streeter-Phelps, (3) full linear dissolved-oxygen balance, and (4) nonlinear dissolved-oxygen balance.

Configuration of Models

The first step in configuring the hydrodynamic and water-quality models for the lower Roanoke River was the development of a spatial grid to represent the study area. The finite-difference grid was established using a common geometry for both models and would serve as the basis for referring information between the two models. In addition, input files were developed for each model, and specific modules were configured by populating parameters that simulated specific physical processes. Data needed to calibrate the EFDC and WASP models were collected between June 1, 2006, and November 7, 2007. Calibration focused on data collected during 2006 because an initial assessment showed dissolved-oxygen levels to be lower in 2006 than in 2007. The simulation period used for calibration purposes was March 1, 2006–October 31, 2006. Several weeks were allotted for the models to achieve numerical stability (spin-up time), including 3 weeks for EFDC and 1 additional week for WASP, thereby shortening the simulation period available for analysis to April 1, 2006–October 31, 2006. The following sections describe the setup and calibration processes for each model.

Numerical Finite-Difference Grid

A three-dimensional, orthogonal rectangular grid was developed for the lower Roanoke River and flood plain by the EFDC support team with USEPA, Region 4. The EFDC model extended approximately 200 km (124 miles) from the USGS station at Roanoke River at Roanoke Rapids to the station at Roanoke River at NC 45 near Westover, N.C. Grid cells were relatively large (1 km wide) in the flood plain because of the absence of strong lateral gradients for conditions other than water depth. Main channel cells were 1 km long and varied in width between 26 and 400 meters. The EFDC model uses a z-grid coordinate system, for which the number of layers varies throughout the model domain. The main channel grid cells were represented with two layers, and flood-plain grid cells had one layer. For cells with multiple layers, layer thickness was set up as half of the total depth.

Light detection and ranging (LiDAR) data with a vertical accuracy of about 25 centimeters (North Carolina Division of Emergency Management, 2002) were used to develop high-resolution digital elevation models (DEMs). For each grid cell, an elevation value was computed by averaging the higher resolution data over the area of the grid. As a result, a detailed representation of flood-plain topography was available, including hydrologic features such as Devils Gut (fig. 2). Channel cross-section surveys performed between 2006 and 2010 were used to develop channel geometry (Wehmeyer and Wagner, 2011). The resulting grid (fig. 3) had 1,847 cells: 1,623 in the flood plain, and 224 in the channel. The grid encompassed approximately 1,692 km², almost half the total drainage area of the lower Roanoke River Basin.

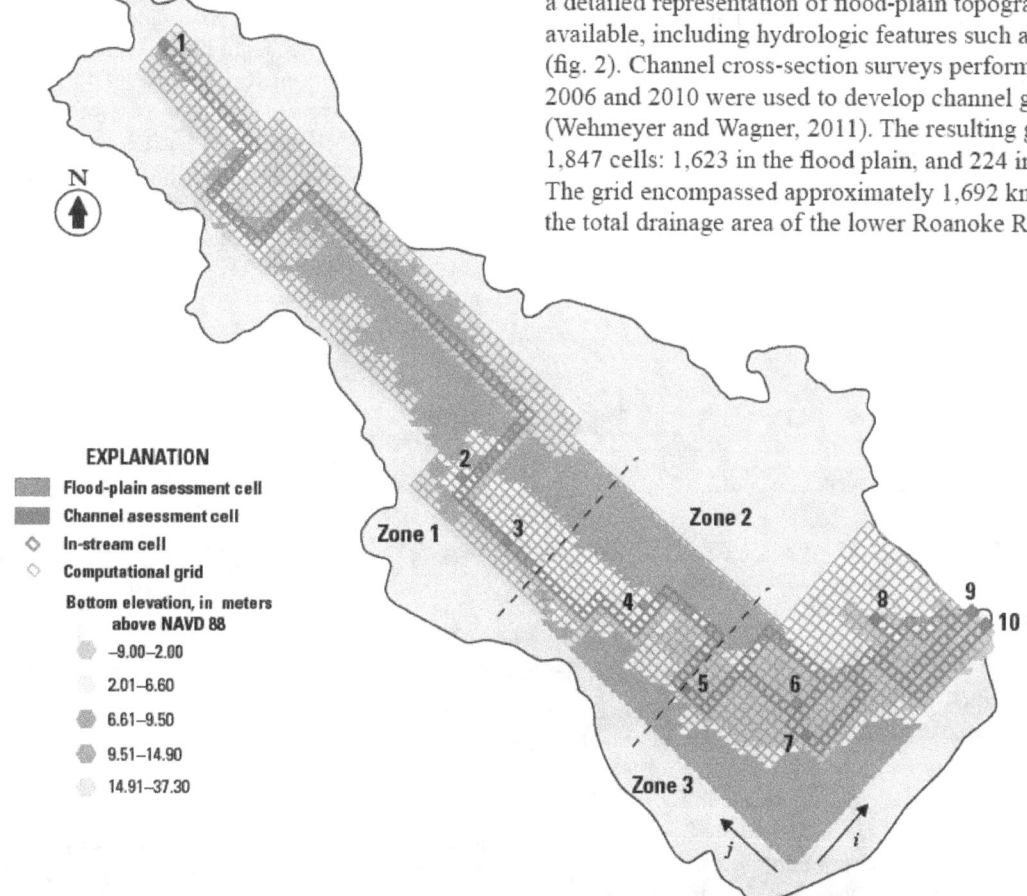

Figure 3. Numerical computational grid and locations of upstream boundary, downstream boundary, and calibration cells. Bottom elevations, in meters above the datum, are referenced to the North American Vertical Datum of 1988 (NAVD 88). Channel elements are narrower than they appear.

The computational grid was georeferenced to spatial attributes of the lower Roanoke River, and individual cells were linked to water-quality and streamgaging stations (table 3). Although both models (EFDC and WASP) used the same grid, identification of cells differed for each model. For EFDC, cells were identified by the i, j notation of a coordinate system with origin at the southwesternmost cell (fig. 3). For WASP, cells were identified with a sequential numbering system. Using the i, j notation, boundary cells for both models were cells 17,110; 26,14; 33,5; and 33,7. Three cells in the main channel were used for calibration, and three cells in the flood plain were monitored to evaluate the effects of flood-plain processes on in-stream dissolved-oxygen levels (table 3).

Table 3. List of channel and flood-plain cells used for evaluating model performance and scenarios. For EFDC, mapping refers to an i, j coordinate system centered at the southwesternmost cell. For WASP, mapping refers to a set of sequential numbers in ascending j-value.

Identi-fier from figure 2	Location represented by grid cell	EFDC mapping		WASP mapping
		i	j	Cell Number
1	Roanoke River at Roanoke Rapids	17	110	1,843
2	Floodplain near Kehukee Swamp	7	57	1,204
3	Roanoke River near Oak City	5	49	1,048
4	Floodplain near Broadneck Swamp	11	34	784
5	Roanoke River at Williamston	7	25	618
6	Floodplain near Devil's Gut	14	15	323
7	Roanoke River at Jamesville	12	12	259
8	Cashie River at San Souci Ferry	26	14	335
9	Cashie River near NC-45 Bridge	33	5	186
10	Roanoke River at NC-45 Bridge	33	7	154

To facilitate calibrating spatially variable model parameters, model cells were classified as belonging to one of three zones (fig. 3). Zone 1 is the drainage area upstream and near the Oak City cell location (number 3 in fig. 3). This portion of the river is relatively incised, and the flood-plain vegetation is representative of bottomland forest. Flows near the USGS streamgaging station near Oak City are primarily one-dimensional, and most of the water released from Roanoke Rapids Dam is conveyed (Lebo, 1998) in the main channel. Zone 2 encompasses the channel and flood-plain cells near Williamston, N.C. This zone is dominated by frequently

inundated flood plains and is representative of hydroperiod conditions necessary for swamp forest. Finally, zone 3 extends to Albemarle Sound, and the flow regime, which is influenced by tides, exhibits substantial lateral gradients in depth. Near constant inundation is characteristic of the peatland forest habitat in zone 3.

EFDC Model Setup and Calibration

The variables simulated with the EFDC model were streamflow, water level, and temperature. The EFDC model was calibrated to measured streamflow and in-stream water level. A single input file (QSER.inp) contained the upstream boundary condition (discharge records from USGS stream-gaging station at Ronaoke River at Roanoke Rapids) and discharge time series of point-source inflows. Inflows from tributaries were not explicitly represented. The model domain is large enough that these contributions were implicitly accounted for by simulating for precipitation and runoff from all grid cells. Furthermore, the largest tributary, the Cashie River, is located in the tidal zone, and flow direction fluctuates such that the net inflow to the Roanoke River is negligible (fig. 4). Time series data for the downstream pressure boundaries (PSER.inp) were developed using water levels measured near NC 45. Continuous temperature measurements made at USGS stations Roanoke River near Halifax and NC 45 were also boundary conditions for EFDC (TSER.inp). An input file for climatic data (ASER.inp) was configured using cloud cover, evaporation, precipitation, air temperature, atmospheric pressure, and solar radiation data measured at a National Climatic Data Center weather station at the Halifax County airport (WBAN 93796)

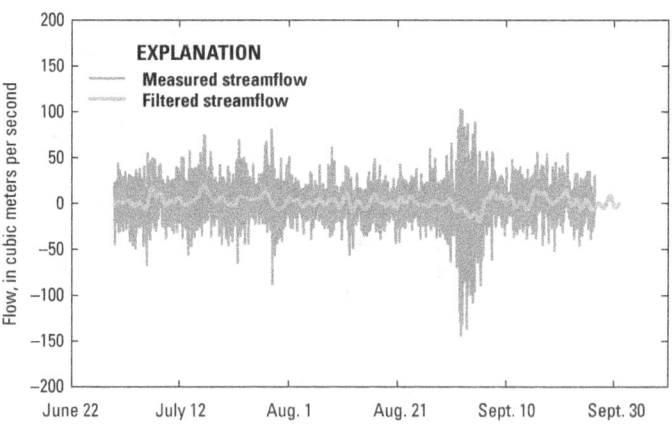

Figure 4. Measured streamflow at U.S. Geological Survey gaging station at Cashie River at Sans Souci Ferry, N.C. (0208113400), and streamflow filtered to remove the effect of tidal backwater (June–September 2006).

Model calibration was approached systematically, the hydrodynamic model was calibrated before the water-quality model, and calibration proceeded from upstream to downstream. Model performance commonly is evaluated using a regression correlation coefficient (R^2) to measure how well the simulated and observed data match. The coefficient can have values between 0 and 1; a value of 0 indicates no correlation, and a value of 1 indicates that the simulated values equal the corresponding measured values. A calibration criterion of 0.5 was used to determine whether the model had explained most of the variability measured.

Calibration was performed first for the Oak City station (figs. 5 and 6) with a focus on parameters that characterize channel conveyance capacity. The detailed description of topography afforded by the use of LiDAR data, led to a well calibrated initial model, and further hydrodynamic calibration was performed by adjusting only a few parameters. The roughness coefficients for the channel and flood plain were adjusted to account for the effect of existing flood-plain vegetation on stream velocities. Final calibrated values ranged between 0.03 and 0.1, with higher values associated with the flood plain (table 4). Simulated hourly flows and water levels were in good agreement with observed measurements as reflected by R^2 values (0.91 for flows and 0.9 for water levels), which were above the calibration criterion value of 0.5. Peak streamflows were underpredicted, although water levels were well simulated. The model deviated from streamflow measurements for specific events. For example, a streamflow peak in June 2006 was not reflected in the boundary streamflow at Roanoke Rapids but was present in the streamflow measurements at Oak City. It is possible that this event may represent a storm not included in the measured precipitation data.

Table 4.   List of calibrated parameters in the EFDC and WASP models.

[g, grams; m², square meters; SOD, sediment oxygen demand; BOD, biochemical oxygen demand]

Parameter	Parameter name	Units	Model	Value	
ZROUGH	Log law roughness height	meters	EFDC	0.1	Flood plain
				0.03	Channel
NDRYSTP	Minimum number of timesteps a cell remains dry after initial drying	dimensionless	EFDC	16	
HDRY	Depth at which cell becomes dry	meters	EFDC	0.03	
HWET	Depth at which cell becomes wet	meters	EFDC	0.08	
SOD	Sediment oxygen demand	g/m²/day	WASP	0	Channel cells
				3	Zone 1, flood-plain cells
				5	Zone 2, flood-plain cells
				5	Zone 3, flood-plain cells
SODT	SOD temperature correction factor	dimensionless	WASP	1.01	Flood-plain cells, all zones
CBOD1	BOD decay rate constant at 20 degrees Celsius	per day	WASP	0.04	
CBOD1T	BOD temperature correction factor	dimensionless	WASP	1.05	

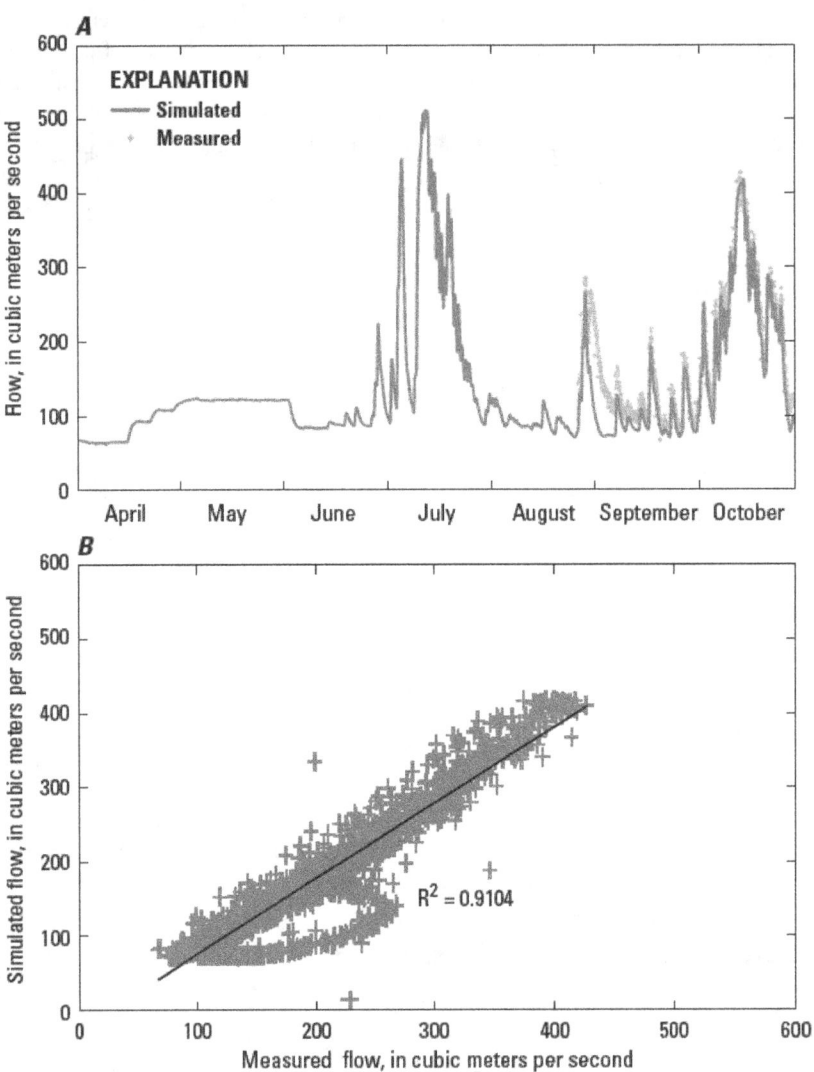

Figure 5.   The Environmental Fluid Dynamics Code (EFDC) model calibration results for Roanoke River at Oak City, North Carolina. *A,* Hourly simulated and measured flows (April–October 2006). *B,* Hourly simulated flow related to measured flow and best-fit line.

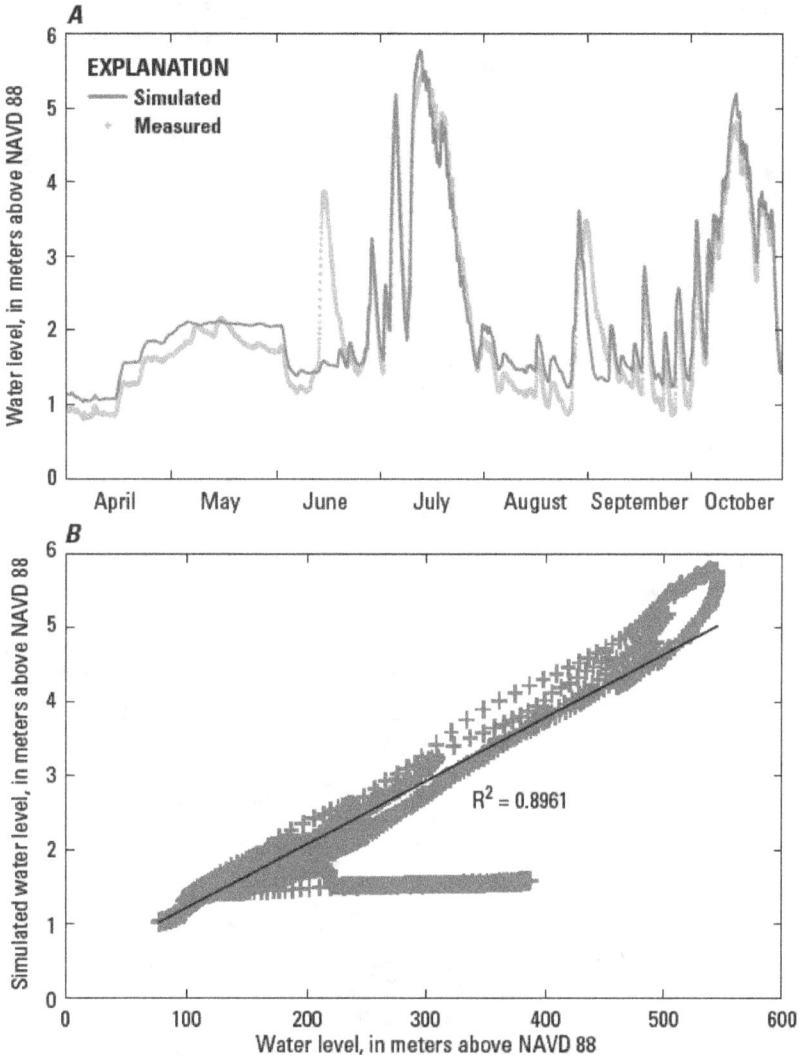

Figure 6.   The Environmental Fluid Dynamics Code (EFDC) model calibration results for Roanoke River at Oak City, North Carolina. *A*, Hourly simulated and measured water levels, in meters above NAVD 88 (April–October 2006). *B*, Hourly simulated water-surface elevation related to measured water-surface elevation and best-fit line.

Given the increased frequency of flood-plain inundation within the river reach between Oak City and Williamston, which roughly corresponds with zone 2 grid cells, a greater degree of uncertainty is expected compared to model simulation of zone 1. Flow paths near Williamston are two-dimensional, whereas flows upstream from Oak City are predominantly one-dimensional. Although streamflow and water levels are well simulated, model accuracy at the USGS station Roanoke River at Williamston decreased slightly (figs. 7 and 8), with an R^2 value of 0.85 for stream-flow and 0.83 for water levels. Parameters that are related to flood-plain processes were adjusted during the calibration in zone 2. These parameters include the threshold levels of cell wetting and drying, which are necessary for model stability (table 4).

A notable source of error for this calibration was a systematic overprediction of surface-water levels (June and September shown in fig. 8). No such discrepancies were present for the Oak City calibration (fig. 6), which suggests limitations in the representation of flood-plain processes. The presence of extensive cuts and gullies was probably not accurately represented by the square-kilometer cells in the model domain. These features are shortcuts for the flow that quickly convey water across different segments of the lower Roanoke River and, as a result, could dampen the shape of the flood wave downstream.

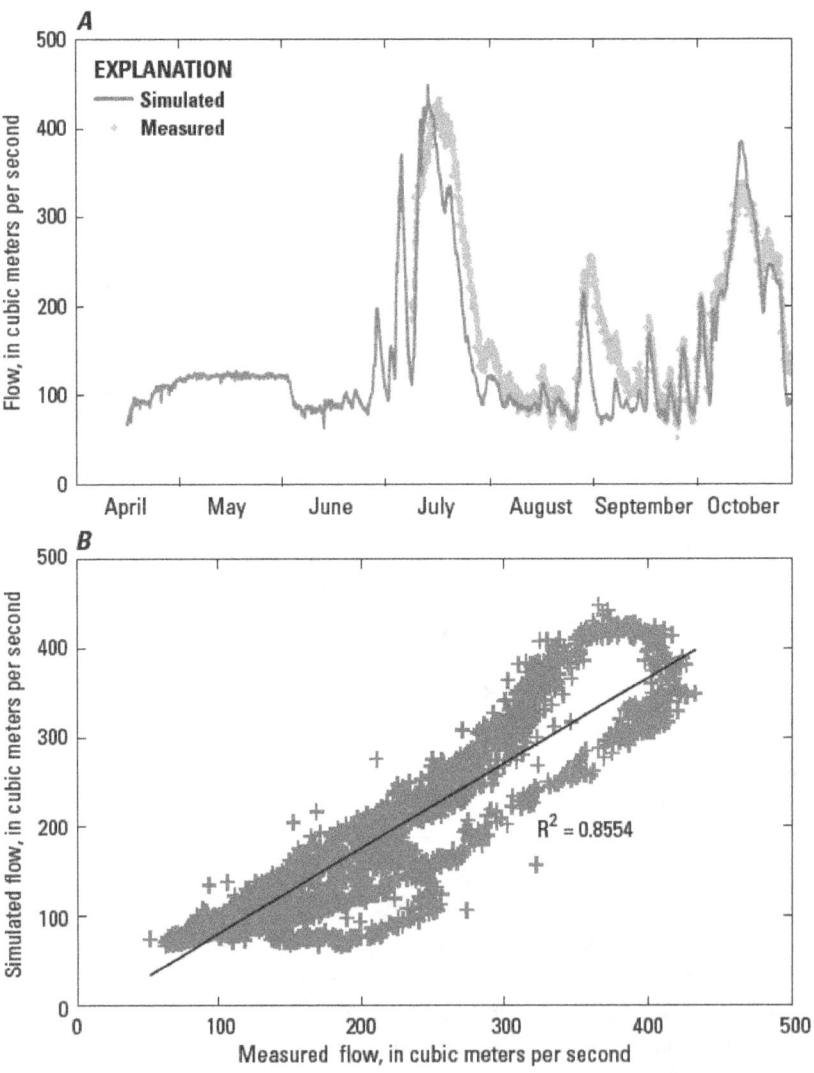

Figure 7.   The Environmental Fluid Dynamics Code (EFDC) model calibration results for Roanoke River at Williamston, North Carolina. *A,* Hourly simulated and measured flows (April–October 2006). *B,* Hourly simulated flow related to measured flow and best-fit line.

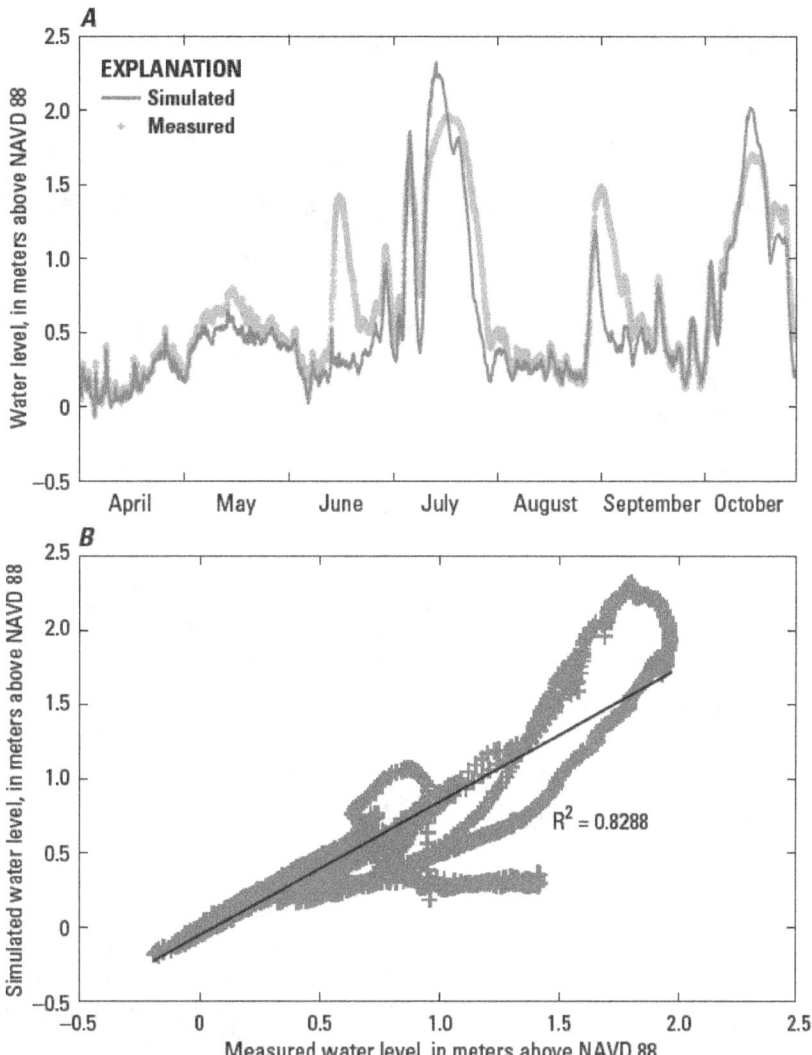

Figure 8. The Environmental Fluid Dynamics Code (EFDC) model calibration results for Roanoke River at Williamston, North Carolina. *A*, Hourly simulated and measured water levels, in meters above NAVD 88 (April–October 2006). *B*, Hourly simulated water-surface elevation related to measured water-surface elevation and best-fit line.

Flood-plain processes appreciably affect the dissolved-oxygen dynamics of the lower Roanoke River. Thus, an effort was made to ensure a reasonable representation of flood-plain hydrology. Water levels in the flood plain measured at a location near Broadneck Swamp were compared to simulation results for an equivalent grid cell (fig. 9). Measured flood-plain levels were reasonably simulated; however, the peak level simulated in July was lower than the measured peak by approximately 0.6 meters (fig. 9). This result strengthened the hypothesis that model limitations are mostly attributed to inaccuracies in geometry and that the conveyance capacity of the flood plain was underpredicted.

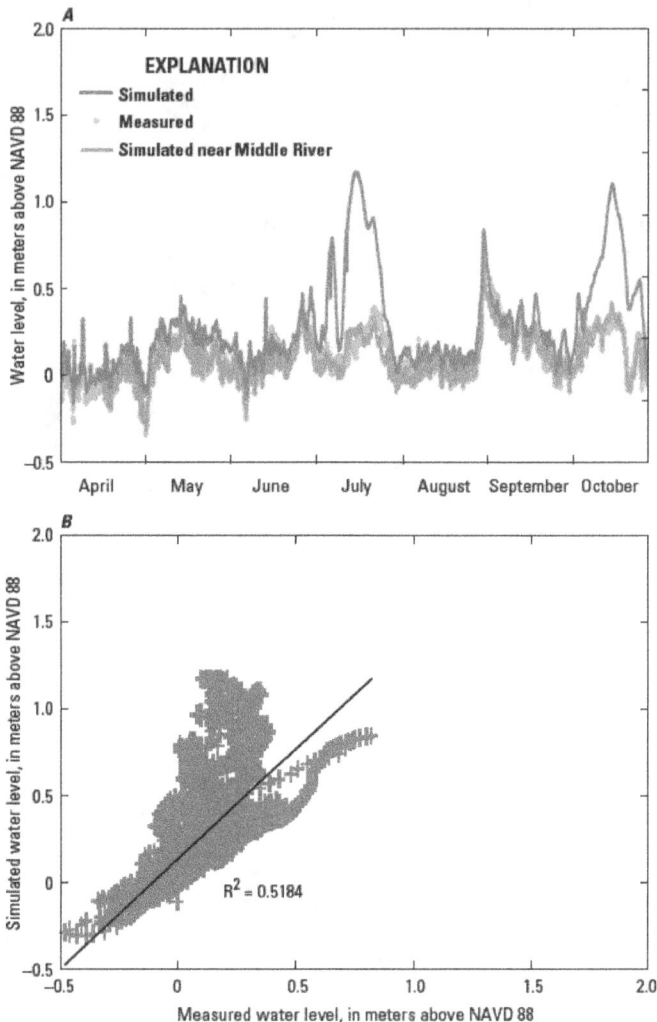

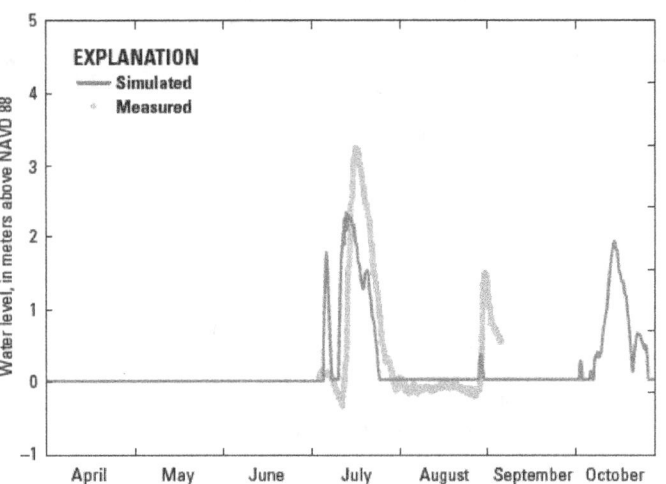

Figure 9.   The Environmental Fluid Dynamics Code (EFDC) model calibration results for the flood plain near Broadneck Swamp of hourly simulated and measured water levels in meters above NAVD 88 (April–October 2006).

Figure 10.   The Environmental Fluid Dynamics Code (EFDC) validation for Roanoke River at Jamesville, North Carolina. *A,* Hourly simulated and measured water levels, in meters above NAVD 88 (April–October 2006). *B,* Hourly simulated water-surface elevation related to measured water-surface elevation and best-fit line.

Measured and simulated water levels at Jamesville were compared for model validation (fig. 10). Water levels near this location (zone 3 grid cells) were greatly influenced by the tide in Batchelor Bay. Calibration of water levels at Jamesville presented challenges, and the final model predictions had an R^2 of 0.52, which was just above the calibration criterion. Some appreciable limitations were noted. Overprediction of high water levels, mentioned in the discussion of calibration at Williamston, was more notable. It is possible that these peaks represent water volume conveyed and diffused by interaction with pathways in the flood plain not represented in the model. Simulated water levels at a grid cell downstream from Jamesville, near the Middle River confluence, compared favorably with measured water levels at Jamesville and did not present these overpredictions probably because of the increased representation of bifurcations and channel splits in the model downstream from Jamesville.

The EFDC model was used to simulate water temperature, but calibration was not performed. A comparison of measured and simulated average temperatures at Williamston (fig. 11) showed that the model slightly overpredicted temperatures, especially during the summer, which led to the systematic error of simulating higher dissolved-oxygen concentrations. When comparing results across the model scenarios, however, the relative differences are unaffected by this limitation.

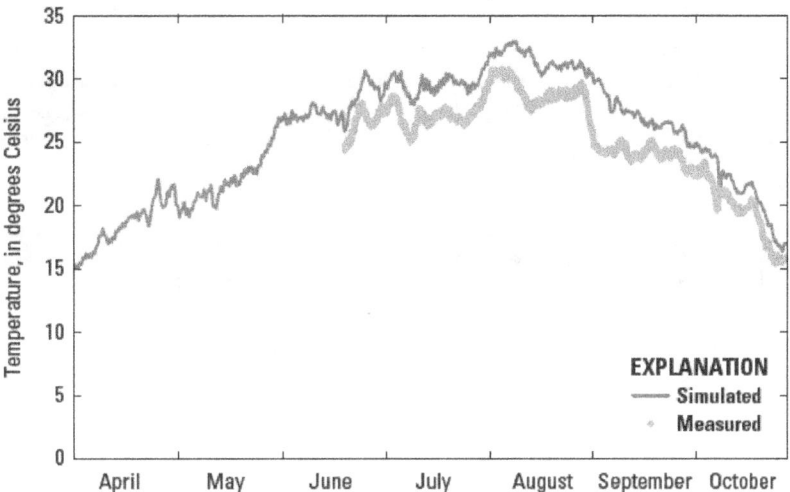

Figure 11. The Environmental Fluid Dynamics Code (EFDC) validation for Roanoke River at Williamston, North Carolina, of hourly simulated and measured water temperatures (April–October 2006).

WASP Setup and Calibration

The WASP model was configured to use streamflow, water-level, and temperature data generated from EFDC model output to simulate dissolved-oxygen concentrations. Spatial segmentation and simulation time periods were the same as those used for the EFDC model. The eutrophication submodel was configured for a simple Streeter-Phelps analysis in which carbonaceous BOD, sediment oxygen demand (SOD), and reaeration were simulated, and calculations for the nitrogen, phosphorus, and phytoplankton variables were bypassed. To simulate reaeration, the Owens reaeration equation was used in which the rate of oxygen transfer is dependent on stream velocity and depth. Time series of measured dissolved oxygen and BOD at USGS stations Roanoke River near Roanoke Rapids and NC 45 were used as upstream and downstream boundary conditions, respectively.

Point sources were simulated by incorporating time series of dissolved-oxygen concentrations in effluent. In addition, measurements of BOD were used in the model where available. The WASP model was configured to track BOD from point sources and ambient BOD from the water column. An ambient BOD of 2 milligrams per liter (mg/L; Bales and Walters, 2003) was set for upstream inflows and as initial conditions for water in all segments of the model domain.

Sediment oxygen demand rate was set to vary across the three zones for flood-plain cells. Initial values were based on a study (Todd and others, 2009) of in-stream swamps in southern Georgia that reported elevated SOD from 0.5 to 14.2 grams of oxygen per square meter per day for coastal plain swamps. Final calibrated values for the Roanoke River flood plain ranged between 1.2 and 5.0 grams of oxygen per square meter per day. A temperature factor of 1.01 was associated with SOD to account for the change in rates with temperature.

The water-quality calibration consisted of running the hydrodynamic model, linking the output to WASP, running the water-quality model, and comparing dissolved-oxygen time series output to available measured data. Calibration was considered successful if the R^2 of dissolved-oxygen concentrations compared favorably to other reported calibrations, including the use of WASP for the Neuse River, where R^2 varied between 0.5 and 0.6 (Wool and others, 2003).

Dissolved-oxygen levels compared well with monitored data at the Oak City location (fig. 12). The regression R^2 obtained was 0.8, which is within the calibration criterion. Discrepancies are noted in June and July, which coincide with periods when the EFDC model also presented discrepancies relative to the measured hydraulic conditions. The large dip in simulated dissolved oxygen that occurred in June coincides with a measured spike in streamflow that was not modeled and may have been a storm event not accounted for in the climate data.

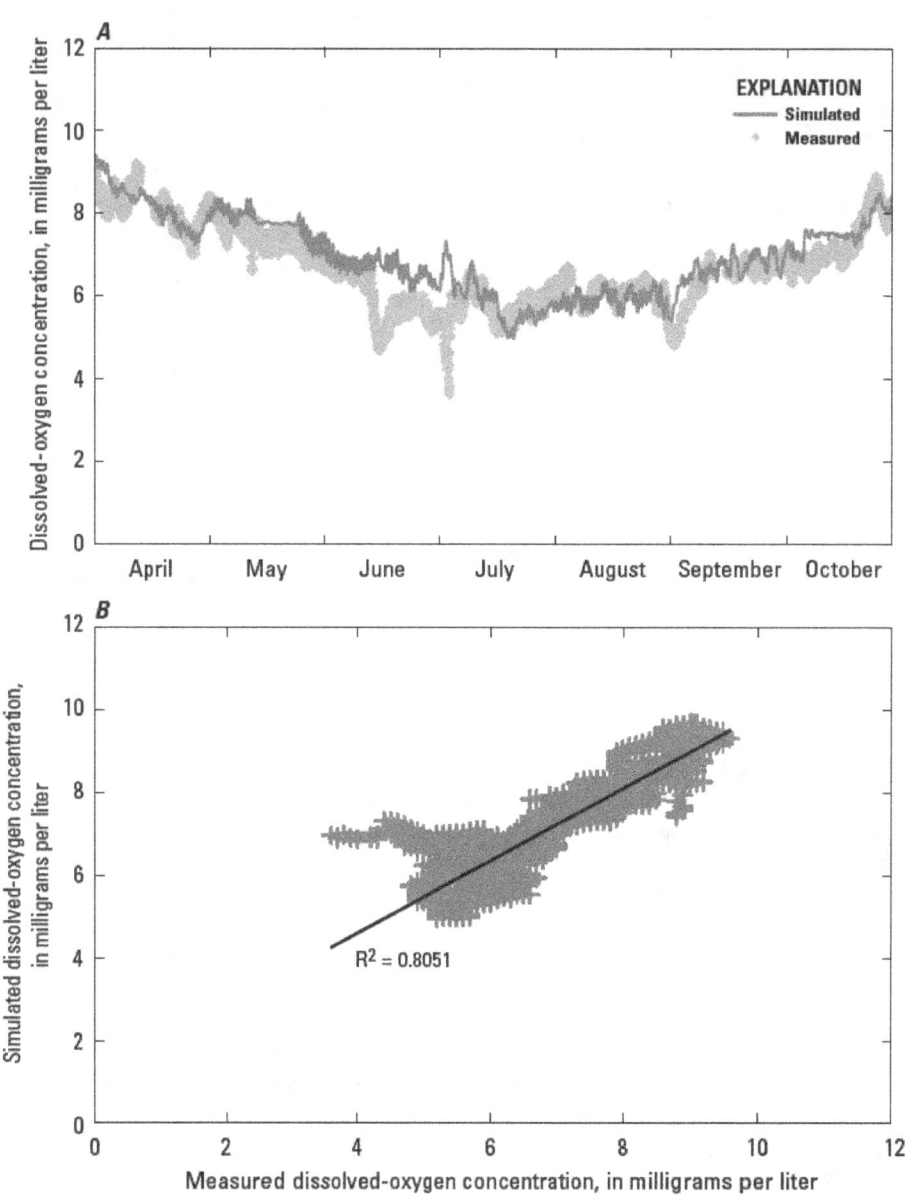

Figure 12. The Water Quality Analysis Simulation Program (WASP) calibration for Roanoke River at Oak City, North Carolina. *A*, Hourly simulated and measured dissolved-oxygen concentrations (April–October 2006). *B*, Hourly simulated dissolved-oxygen concentrations related to measured dissolved-oxygen concentrations and best-fit line.

At the Williamston location simulated and measured dissolved-oxygen concentrations also compared well (fig. 13). Uncertainties in flood-plain dynamics carried over to introduce errors in dissolved-oxygen levels, especially at low flows. Calibration focused on SOD rates such that minimum values of dissolved-oxygen concentrations attributable to flood-plain drainage could be captured. The plot in figure 13, which corresponds to model results without simulating the flood plain, illustrates the effect of the flood plain on dissolved-oxygen levels. Minimum concentrations in particular were greatly affected by flood-plain drainage. When comparing simulations of dissolved oxygen between model runs that include and do not include the flood plain, minimum

concentrations are 61 percent lower at the Williamston site and 67 percent lower at the Jamesville site in model simulations that account for the flood-plain interaction. A dissolved-oxygen decline that occurred about July 21 (fig. 13) represents a 22 percent dissolved-oxygen decrease over 10 days. Without the flood-plain interaction, model simulations predict a 9-percent decrease for the same interval. It was hypothesized that the difference was attributable to flood-plain interaction, and by including the flood-plain simulation, the model predicted a 30-percent decrease. Inaccuracies were the result of not being able to model some of the flood-plain water storage and conveyance, which could increase dilution and reaeration processes.

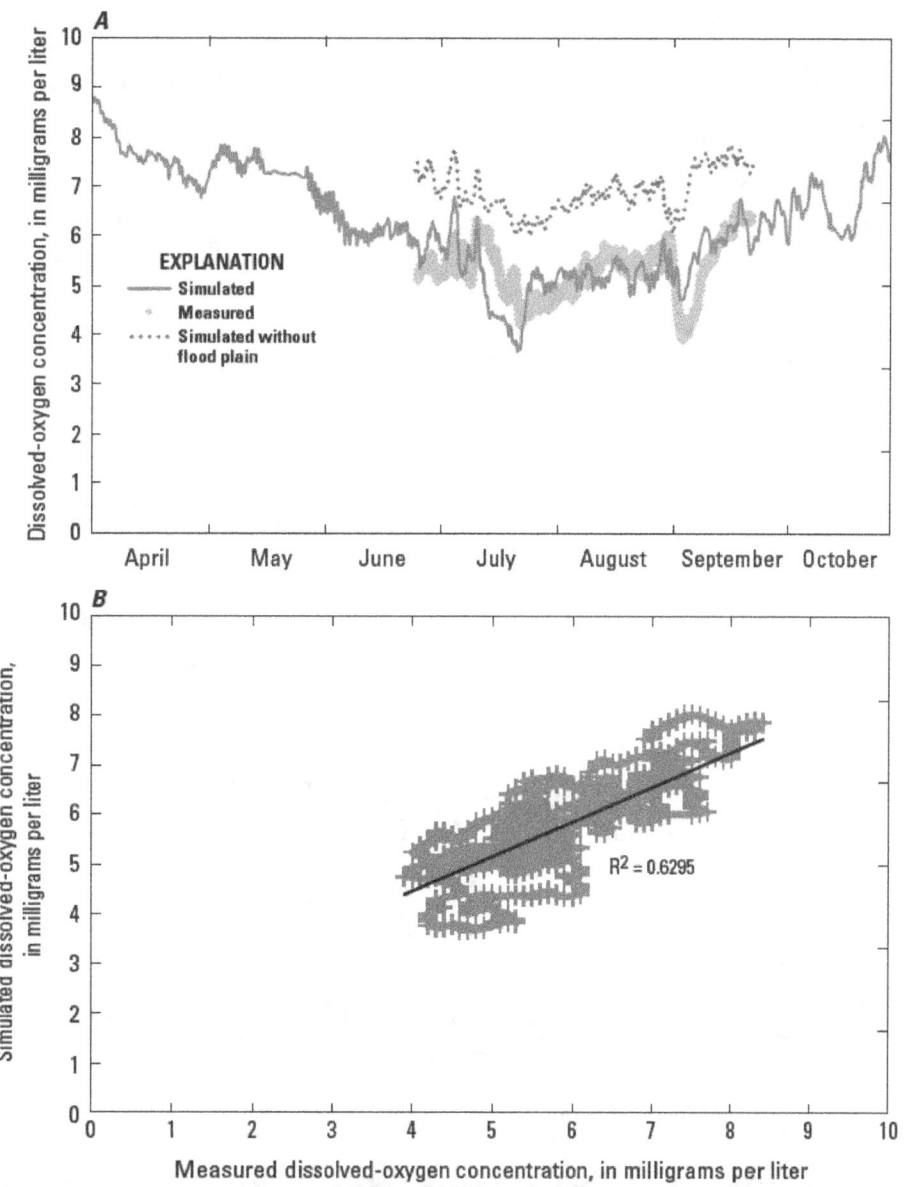

Figure 13. The Water Quality Analysis Simulation Program (WASP) calibration for Roanoke River at Williamston, North Carolina. *A,* Hourly simulated and measured dissolved-oxygen concentrations (April–October 2006). *B,* Hourly simulated dissolved-oxygen concentrations related to measured dissolved-oxygen concentrations and best-fit line.

The effect of flood-plain processes on dissolved oxygen is most pronounced at the Jamesville location (site 8, fig. 2) and in the nearby swamps. The dissolved oxygen 10-day decrease (July 12–22, fig. 14) is 27 percent and is clearly not represented in a model that does not account for the flood plain. For the same time period, the model that simulated the flood plains predicted a 3-percent decrease.

Measured and simulated dissolved-oxygen concentrations were compared at a flood-plain location near Devils Gut.

The range of observed concentrations was represented in the simulated results (fig. 15). It is hypothesized that despite the limitations of the model, the parameterization of dissolved-oxygen processes, including the SOD calibration of the flood plain, led to a reasonably calibrated model applicable to styling the effect of management scenarios on in-stream dissolved-oxygen levels.

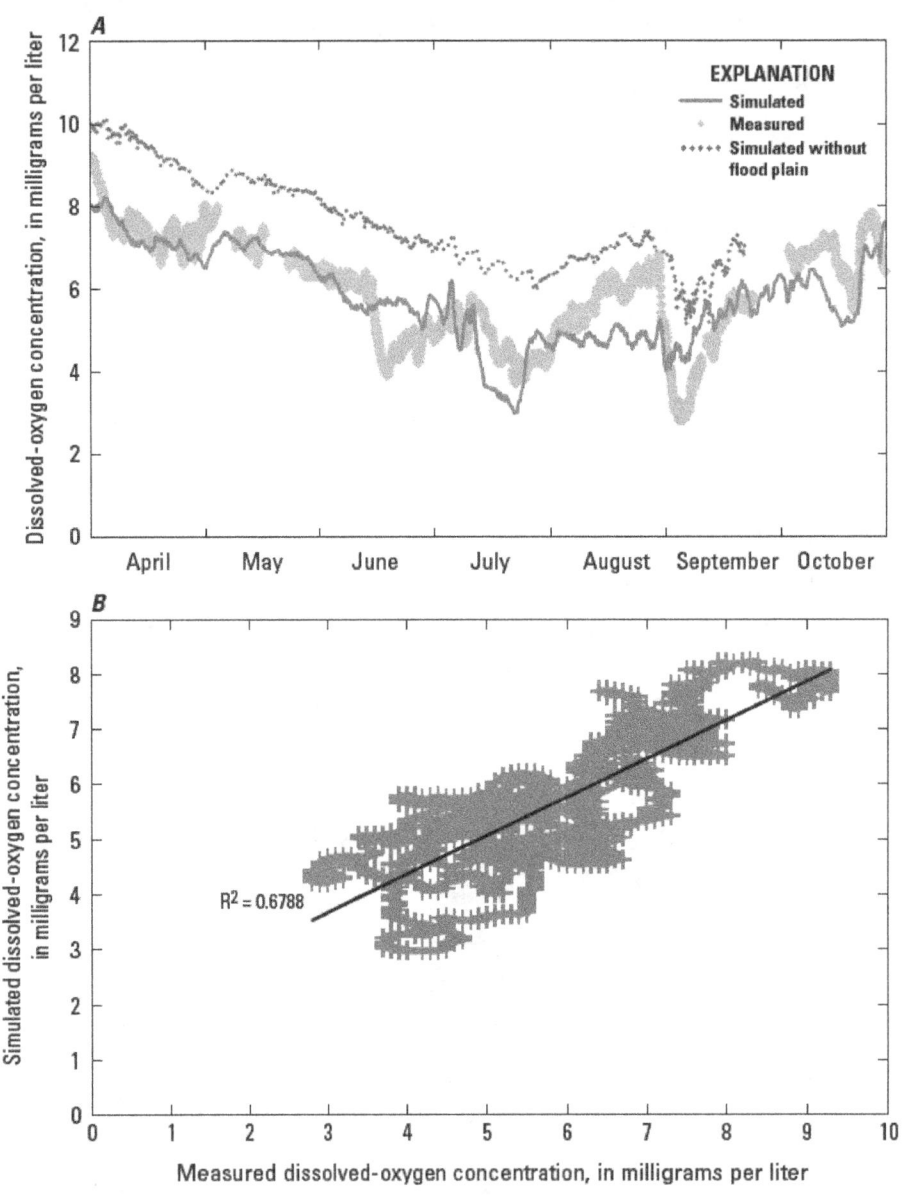

Figure 14. The Water Quality Analysis Simulation Program (WASP) calibration for Roanoke River at Jamesville, North Carolina. *A*, Hourly simulated and measured dissolved-oxygen concentrations (April–October 2006). *B*, Hourly simulated dissolved-oxygen concentrations related to measured dissolved-oxygen concentrations and best-fit line.

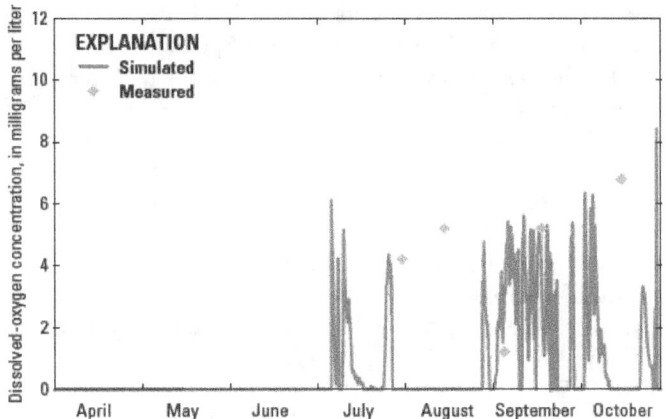

Figure 15. Simulated dissolved-oxygen concentrations at flood-plain grid cell and measured dissolved-oxygen concentrations at a swamp site near Devils Gut (April–October 2006).

Scenario Evaluation

Scenarios to evaluate management options were developed by the U.S. Army Corps of Engineers in collaboration with resource management agencies (table 5).

The scenarios included a set of flow release schedules for Kerr Dam and improvements to dissolved-oxygen levels at Roanoke Rapids Dam. The effects of hydropower peaking operations were also included as a scenario.

Scenarios were evaluated by applying the EFDC/WASP modeling system for the Roanoke River under varying release schedules. Boundary conditions, in the form of streamflow at Roanoke Rapids Dam were provided by the USACOE for each modification to the Kerr Dam release schedule. Simulation results for each scenario were evaluated at six locations on the lower Roanoke River, including the three main-stem locations used in the calibration of the model: Roanoke River at Oak City, Roanoke River at Williamston, and Roanoke River at Jamesville. Riparian or flood-plain water quality was assessed at three locations Kehukee Swamp, which is located near Oak City, N.C., Broadneck Swamp located near Williamston, N.C., and Devils Gut, which is near the Jamesville site (figs. 2, 3; table 3).

The scenarios were constructed using input data collected during the spring and summer of 2006. During this period, several factors (temperature, streamflow, and so forth) combined to create low dissolved-oxygen levels in the river and, as such, represent a critical condition useful for evaluating the effects of flow-release scenarios on dissolved oxygen.

Table 5. Scenarios developed by the U.S. Army Corps of Engineers and partners to evaluate management options.

[FERC, Federal Energy Regulatory Commission; DO, dissolved oxygen; mg/L, miligrams per liter; ft³/s, cubic feet per second]

Scenario no.	Scenario name	Scenario description	Baseline	Boundary conditions	
				Flow at dam	Dissolved oxygen
1	Effects of point sources	Assessment of in-stream and flood-plain water quality as a result of removing permitted point-source discharges	Minimum release flows scenario with point sources at permited limits	FERC minimum flows	DO record of 2006
2	Effects of dissolved oxygen improvements	Assessment of in-stream and flood-plain water quality as a result of improving DO concentrations at Roanoke Rapids Dam by 1 mg/L	Minimum release flows scenario with point sources at permited limits	FERC minimum flows	DO record 2006, increased by 1 mg/L
3	Effects of hydropower peaking operations	Assessment of in-stream and flood-plain water quality as a result of hydropower peaking operations	Minimum release flows scenario with no hydropower peaking	FERC minimum flows, with peaking schedule	DO record of 2006
4	Effects of alternative 6b	Assessment of in-stream and flood-plain water quality as a result of implementing alternative 6b	Existing flood-control release flows as implemented without alternative 6b (maximum release 20,000 ft³/s)	Alternative 6b, with maximum flow 35,000 ft³/s	DO record of 2006
5	Evaluation of a dynamic stepping-down release schedule (Betterment Plan)	Assessment of in-stream and flood-plain water quality as a result of implementing the Betterment Plan	Existing flood-control release flows as implemented without the Betterment Plan	Existing operations flood-control flows, with stepping down procedure	DO record of 2006

Effects of Point Sources on Roanoke River Water Quality

Three major industrial facilities and wastewater-treatment plants are permitted to discharge into the Roanoke River. These facilities discharge large volumes of wastewater with high BOD load that may influence near and far field dissolved-oxygen conditions (Bales and Walters, 2003). The calibrated model was used to test the effect of permitted point sources, whereby minimum release flows established by the Federal Energy Regulatory Commission (FERC) were used as upstream boundary conditions at Roanoke Rapids. Downstream flood-plain inundation generally occurs when the weekly average flow is at or above approximately 11,000 ft³/s (312 m³/s) (U.S. Army Corps of Engineers, written commun., 2010). As a result, these minimum-flow scenarios were not affected by flood-plain drainage, thereby isolating the effect of point sources.

In these baseline scenarios, point sources were simulated at fully permitted conditions with point-source discharge dissolved-oxygen concentrations at the daily water-quality standard (5 mg/L). To assess the effect of point sources, a simulation was also performed in which the point sources were removed.

Model predictions showing the effect of eliminating permitted point sources on dissolved-oxygen concentrations are presented in figure 16 and summary statistics are presented in table 6. At Oak City, average dissolved-oxygen concentrations did not change when all upstream point sources were removed, and dissolved-oxygen concentrations at Williamston and Jamesville increased by less than 1 percent. Most of the effect was during the late summer, when temperatures are high. At Williamston, the increase in dissolved-oxygen levels results in a 2-percent reduction in the number of instances dissolved oxygen is below 5 mg/L. At Jamesville, the largest effect was in the reduction of instances that dissolved oxygen is 4 mg/L (table 6). In the model, this site is downstream from the largest wastewater discharger in the study area, which has an average BOD load of approximately 42,000 kilograms.

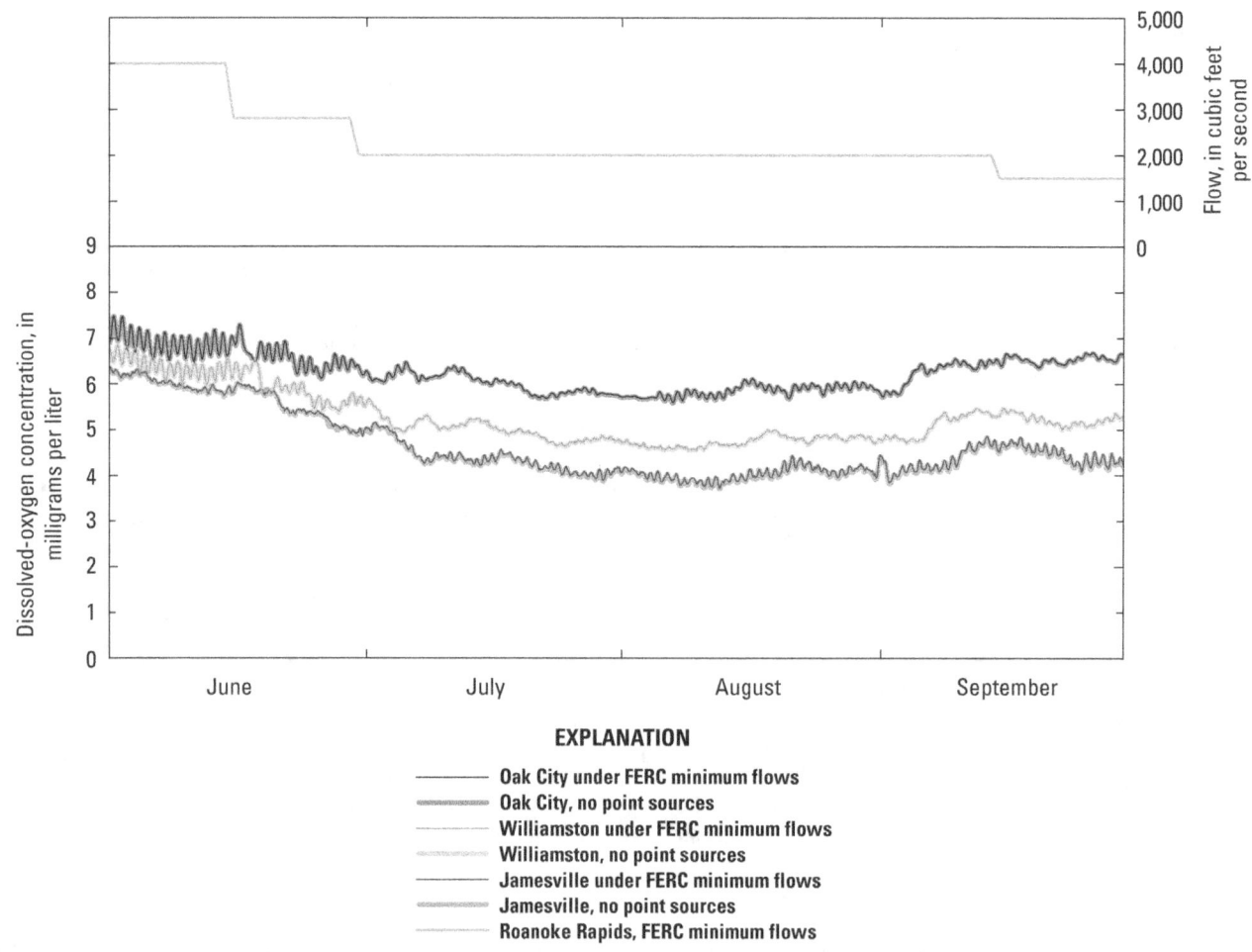

EXPLANATION

———— Oak City under FERC minimum flows
———— Oak City, no point sources
———— Williamston under FERC minimum flows
———— Williamston, no point sources
———— Jamesville under FERC minimum flows
———— Jamesville, no point sources
———— Roanoke Rapids, FERC minimum flows

Figure 16.   Simulated dissolved-oxygen concentrations at Roanoke River near Oak City, Williamston, and Jamesville, North Carolina, under minimum flows and no point sources [FERC, Federal Energy Regulatory Commission].

Table 6. Summary statistics of simulated dissolved-oxygen concentrations and exceedances of reference dissolved-oxygen levels at three locations on the Roanoke River for the effects of point sources scenario.

[mg/L, milligrams per liter. Comparisons were done using instantaneous hourly values and using the Federal Energy Regulatory Commission (FERC) minimum flows scenario as a baseline]

	Maximum	Minimum	Mean	Median	Percentage of time dissolved-oxygen concentration is less than	
					5 mg/L[1]	4 mg/L[2]
Oak City						
FERC mininum flows	7.5	5.5	6.2	6.2	0	0
No point sources	7.5	5.6	6.2	6.2	0	0
Williamston						
FERC mininum flows	6.8	4.6	5.3	5.1	42	0
No point sources	6.8	4.5	5.2	5.1	43	0
Jamesville						
FERC mininum flows	6.4	3.7	4.6	4.4	74	12
No point sources	6.3	3.7	4.6	4.3	74	15

[1] North Carolina water-quality daily standard for dissolved oxygen.

[2] North Carolina water-quality instantaneous standard for dissolved oxygen.

Effects of Variation in Upstream Dissolved-Oxygen Levels on Roanoke River Water Quality

The quality of water discharged from Roanoke Rapids Dam can vary depending on operations and management actions at Kerr Dam. Measures being considered to improve water quality in the lower Roanoke River include structural changes at the upstream dams that could improve dissolved-oxygen levels. To isolate and evaluate the effect of dissolved-oxygen improvements at Roanoke Rapids Dam, a scenario was constructed in which measured dissolved-oxygen levels were increased by 1 mg/L relative to 2006 measured values at Roanoke Rapids. To run this scenario, minimum flow releases from Roanoke Rapids Dam, established by the FERC, were used as inflow model boundary conditions. As mentioned previously, downstream flood-plain inundation generally occurs when the weekly average flow is at or above approximately 11,000 ft³/s (312 m³/s). Therefore, these minimum-flow scenarios were not affected by flood-plain drainage, thereby isolating the effect of variations in dissolved-oxygen levels from water releases.

Simulations indicate that the effects of these upstream changes are most notable near Roanoke Rapids Dam, and effects diminish downstream (fig. 17; table 7). Although the improvement in dissolved-oxygen concentration was held constant during the simulation period, the dissolved-oxygen load (mass of oxygen added) was higher during high flows, which likely explains the reason for larger relative improvements during June.

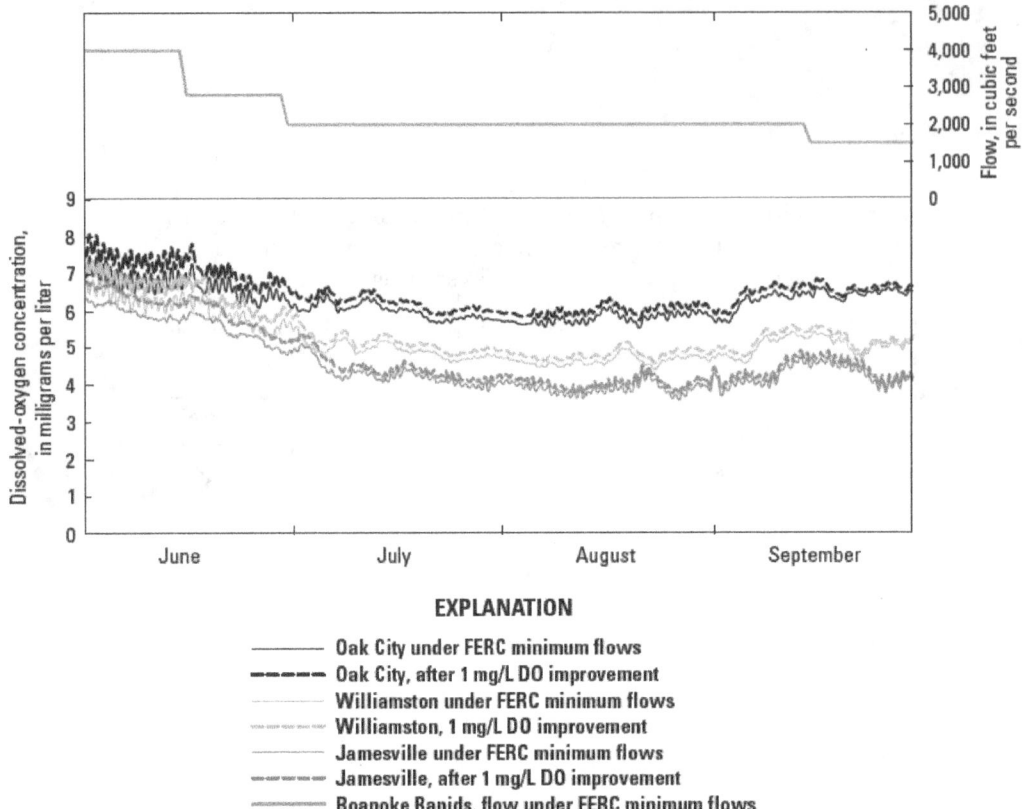

EXPLANATION

	Oak City under FERC minimum flows
------	Oak City, after 1 mg/L DO improvement
------	Williamston under FERC minimum flows
------	Williamston, 1 mg/L DO improvement
------	Jamesville under FERC minimum flows
------	Jamesville, after 1 mg/L DO improvement
------	Roanoke Rapids, flow under FERC minimum flows

Figure 17. Simulated dissolved-oxygen (DO) concentrations at Roanoke River near Oak City, Williamston, and Jamesville, North Carolina, under minimum flows and 1 milligram per liter (mg/L) improvement of dissolved-oxygen levels at Kerr Dam [FERC, Federal Energy Regulatory Commission].

Table 7. Summary statistics of simulated dissolved-oxygen concentrations and exceedances of reference dissolved-oxygen levels at three locations on the Roanoke River for the dissolved-oxygen improvement scenario.

[mg/L, milligrams per liter. Comparisons were done using instantaneous hourly values and using the Federal Energy Regulatory Commission (FERC) minimum flows scenario as a baseline]

	Maximum	Minimum	Mean	Median	Percentage of time dissolved-oxygen concentration is less than	
					5 mg/L[1]	4 mg/L[2]
Oak City						
FERC mininum flows	7.5	5.5	6.2	6.2	0	0
1 mg/L DO improvement	8.1	5.7	6.5	6.4	0	0
Williamston						
FERC mininum flows	6.8	4.4	5.2	5.0	52	0
1 mg/L DO improvement	7.3	4.6	5.4	5.1	38	0
Jamesville						
FERC mininum flows	6.3	3.6	4.5	4.2	76	29
1 mg/L DO improvement	6.8	3.7	4.7	4.4	71	13

[1] North Carolina water-quality daily standard for dissolved oxygen.

[2] North Carolina water-quality instantaneous standard for dissolved oxygen.

Effects of Hydropower Peaking Operations on Roanoke River Water Quality

To evaluate the effect of hydropower peaking, an upstream boundary condition was constructed to reflect the current schedule for peaking operations. During non-flood and non-striped bass spawning periods, daily power generation is scheduled by the power companies according to their minimum flow requirements. During flood events, power companies are permitted to generate power by hydropower peaking operations, resulting in rapid changes from minimum flow to maximum flow, which was 566 m³/s (20,000 ft³/s) under existing operations. Peaking occurs a few hours a day in the hot months of the summer when load demand is highest (fig. 18); therefore, simulations were performed for July 1–September 30, 2006.

As a result of increased interaction with the flood plain and increased velocity-induced aeration, peaking operations for hydropower have the effect of broadening the range of dissolved-oxygen concentrations compared to operations under FERC minimum flows (figs. 19, 20; table 8). For example, at the Oak City location, maximum dissolved-oxygen concentrations increased by 26 percent and minimum concentrations decreased by 16 percent, whereas the mean concentrations remained unchanged. At Williamston, the number of instances dissolved oxygen was below 5 mg/L decreased by 38 percent under peaking operations, and the minimum concentration at Jamesville decreased by 25 percent from 3.6 mg/L to 2.7 mg/L.

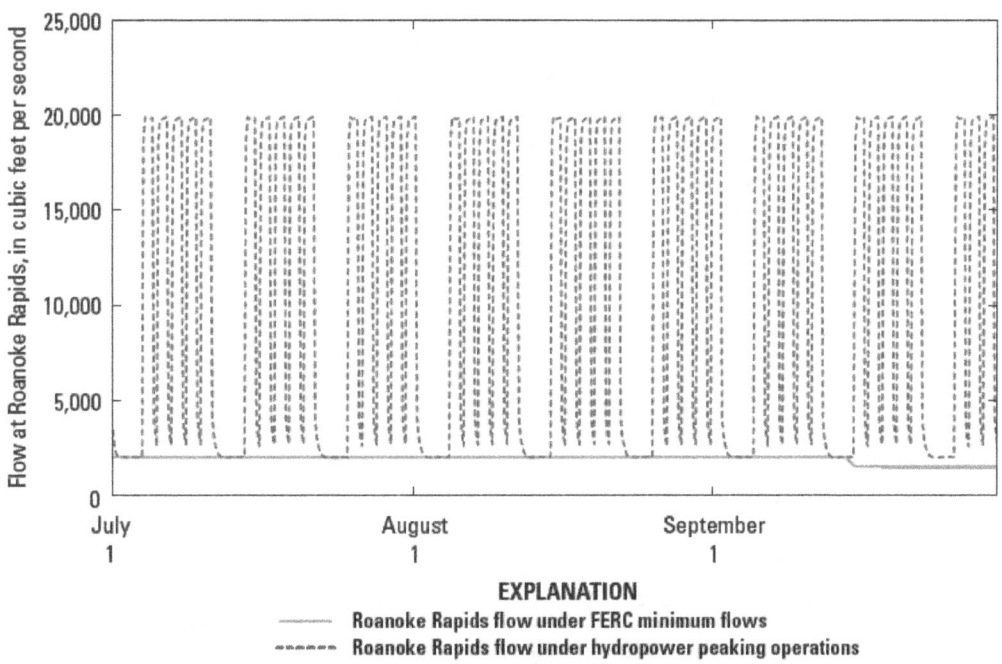

Figure 18. Flow at Roanoke Rapids under Federal Energy Regulatory Commission (FERC) minimum flows and hydropower peaking operations. [Data provided by the U.S. Army Corps of Engineers and Dominion Power, written commun., 2010.]

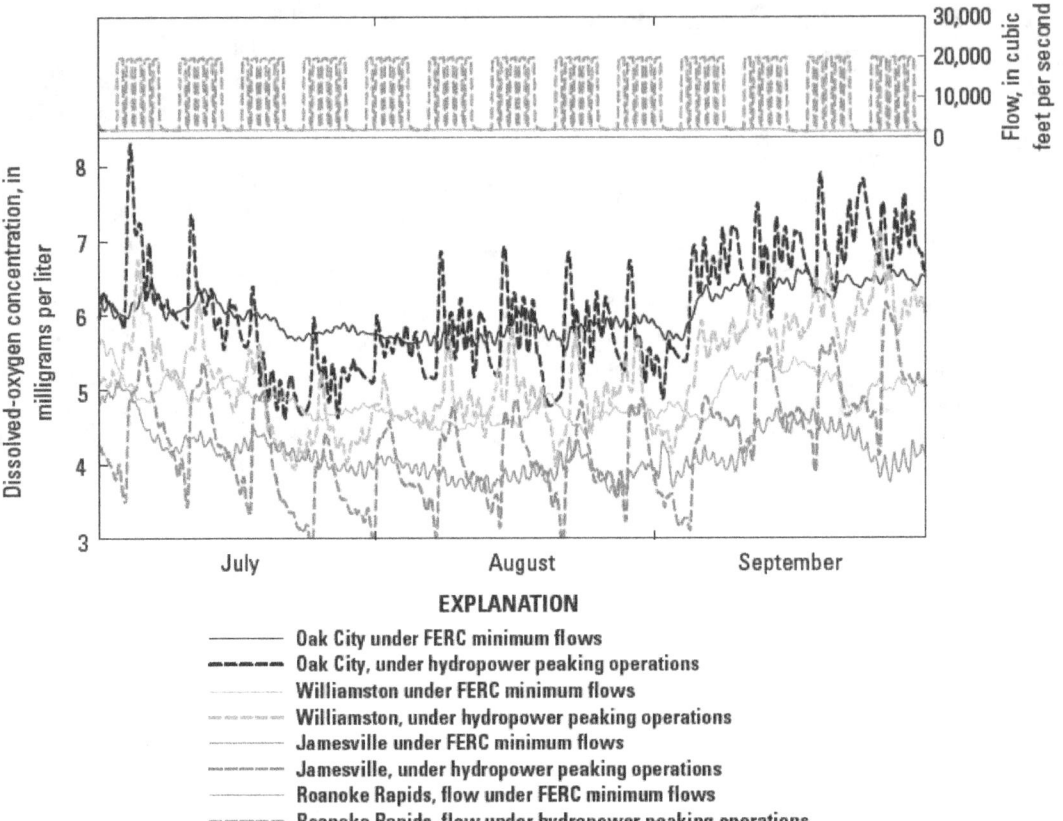

EXPLANATION

——————— Oak City under FERC minimum flows
– – – – – Oak City, under hydropower peaking operations
··········· Williamston under FERC minimum flows
– · – · – Williamston, under hydropower peaking operations
——————— Jamesville under FERC minimum flows
– – – – – Jamesville, under hydropower peaking operations
——————— Roanoke Rapids, flow under FERC minimum flows
– · · – · · – Roanoke Rapids, flow under hydropower peaking operations

Figure 19. Simulated dissolved-oxygen concentrations at Roanoke River near Oak City, Williamston, and Jamesville, North Carolina, under Federal Energy Regulatory Commission (FERC) minimum flows and hydropower peaking.

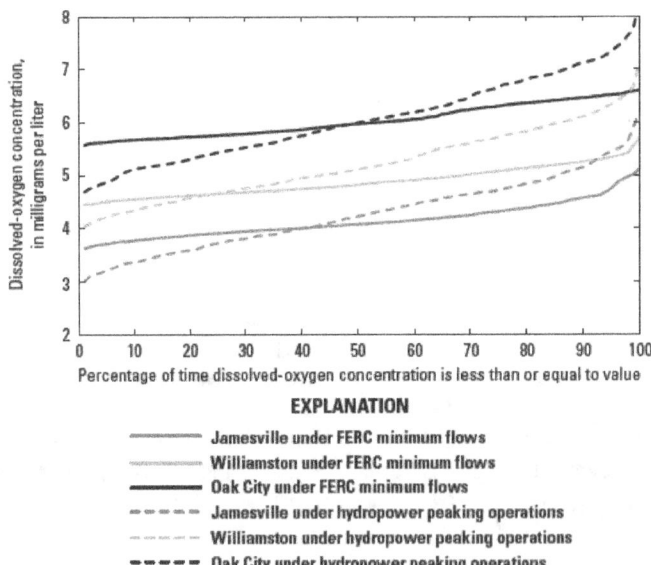

EXPLANATION

——————— Jamesville under FERC minimum flows
··········· Williamston under FERC minimum flows
——————— Oak City under FERC minimum flows
– – – – – Jamesville under hydropower peaking operations
– – – – – Williamston under hydropower peaking operations
– – – – – Oak City under hydropower peaking operations

Figure 20. Frequency of occurrence of dissolved-oxygen concentrations at Roanoke River near Oak City, Williamston, and Jamesville, North Carolina, under Federal Energy Regulatory Commission (FERC) minimum flows and hydropower peaking operations.

Table 8. Summary statistics of simulated dissolved-oxygen concentrations and exceedances of reference dissolved-oxygen levels at three locations on the Roanoke River for the effects of hydropower peaking scenario.

[mg/L, milligrams per liter. Comparisons were done using instantaneous hourly values and using the Federal Energy Regulatory Commission (FERC) minimum flows scenario as a baseline]

	Maxi-mum	Mini-mum	Mean	Median	Percentage of time dissolved-oxygen concentration is less than	
					5 mg/L[1]	4 mg/L[2]
Oak City						
Existing opera-tions	10.0	4.6	6.9	6.8	0	1
Operations under 6b	9.8	4.7	6.9	6.8	0	0
Williamston						
Existing opera-tions	8.0	2.3	4.1	3.7	84	58
Operations under 6b	7.5	2.6	4.5	4.5	72	34
Jamesville						
Existing opera-tions	6.7	1.9	3.4	3.1	86	76
Operations under 6b	6.1	1.7	3.5	3.4	88	70

[1] North Carolina water-quality daily standard for dissolved oxygen

[2] North Carolina water-quality instantaneous standard for dissolved oxygen

The model simulations of the peaking scenario predicted an increased interaction with the flood plain (fig. 21) in the peatland swamp or zone 3 near Jamesville, N.C. Upstream flood-plain locations (Kehukee Swamp and Broadneck Swamp) were not inundated in this scenario. Water levels in the swamp location near Devils Gut oscillated between 0.01 and 1.2 meters, which led to changes in dissolved-oxygen levels between 0.1 and 6.5 mg/L. Minimum dissolved-oxygen concentrations coincided with the low swamp water levels, which have the effect of increasing temperatures and depleting oxygen from the water column. Conversely, the higher dissolved-oxygen concentrations are associated with oxygenation from the influx of water with high dissolved-oxygen concentrations into the flood plain.

Roanoke River Water Quality under Existing Operations and Alternative 6b

An option that is being considered for the management of Kerr Dam is known as alternative 6b (U.S. Army Corps of Engineers, 2010) and would allow the rapid release of water from Kerr Dam flood storage, primarily by releasing up to 35,000 ft³/s when the water level in the reservoir exceeds 303 feet above NAVD 88 (fig. 22). This option would have the effect of inundating more of the downstream flood plain, but for a shorter period than under current conditions, and has been advocated by numerous stakeholders.

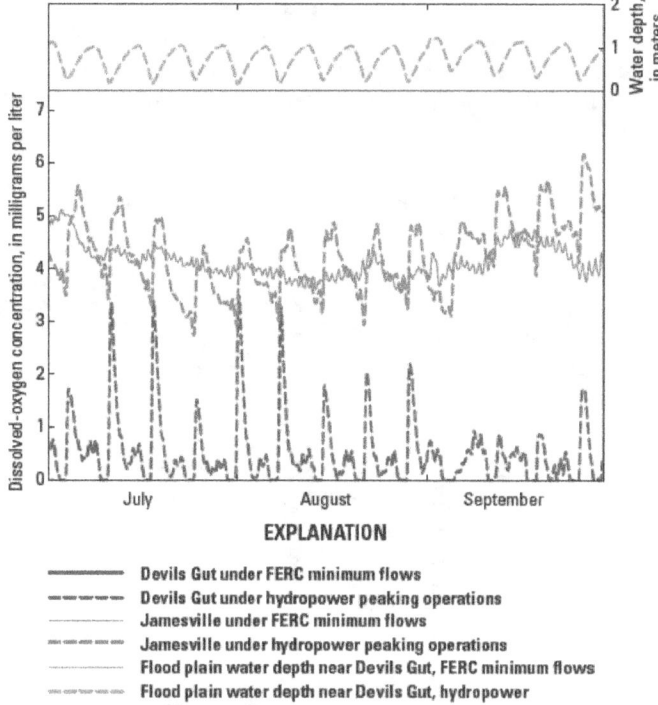

EXPLANATION

——————— Devils Gut under FERC minimum flows
– – – – – Devils Gut under hydropower peaking operations
·············· Jamesville under FERC minimum flows
– ·· – ·· – Jamesville under hydropower peaking operations
·············· Flood plain water depth near Devils Gut, FERC minimum flows
— ·· — ·· — Flood plain water depth near Devils Gut, hydropower peaking operations

Figure 21. Dissolved-oxygen concentrations and water levels at a swamp location near Devils Gut under Federal Energy Regulatory Commission (FERC) minimum flows and hydropower peaking, and dissolved-oxygen concentrations at Roanoke River near Jamesville, North Carolina.

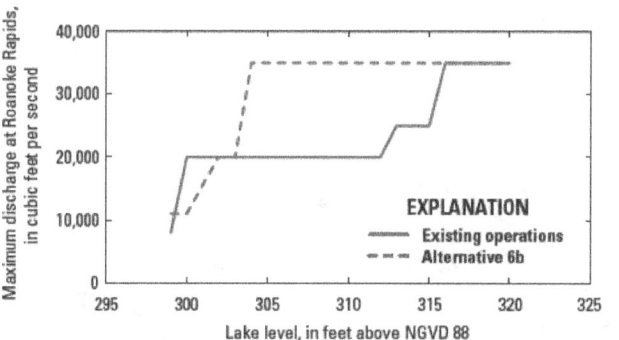

Figure 22. Maximum flow releases at John H. Kerr Dam as a function of reservoir water level for flood management under existing operations and alternative 6b plan for a simulated spring and summer period (July–August). [Adapted from U.S. Army Corps of Engineers, 2010.]

Discharges at Roanoke Rapids for existing operations and alternative 6b are presented in figure 23. These discharges were used as boundary conditions to simulate the effects downstream. Simulated dissolved-oxygen concentrations for the three assessment locations in the main channel of the Roanoke River are presented in figures 24 and 25, and summary statistics for the same locations are presented in table 9.

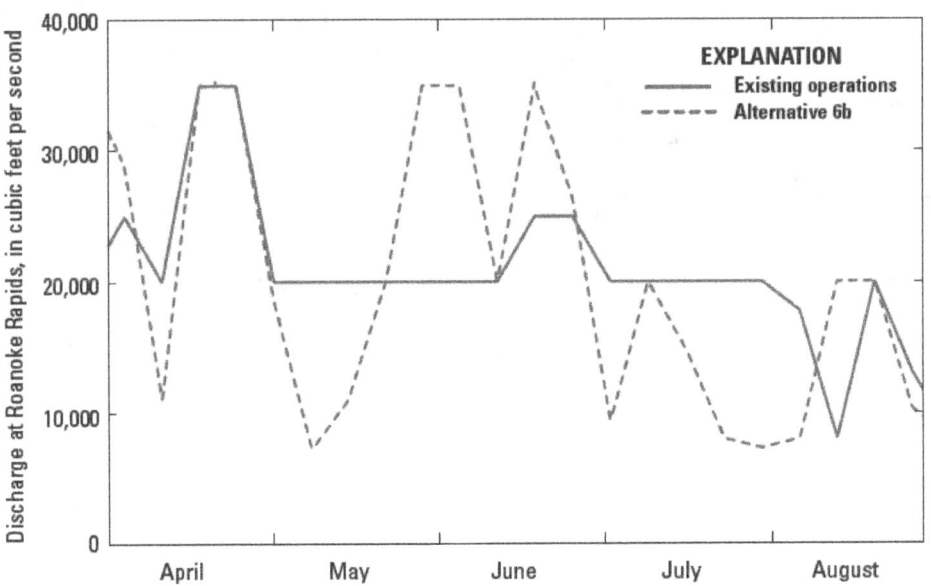

Figure 23.   Weekly discharge at Roanoke Rapids under existing operations and under alternative 6b for a simulated spring and summer period (July –August). [Data provided by the U.S. Army Corps of Engineers, written commun., 2010.]

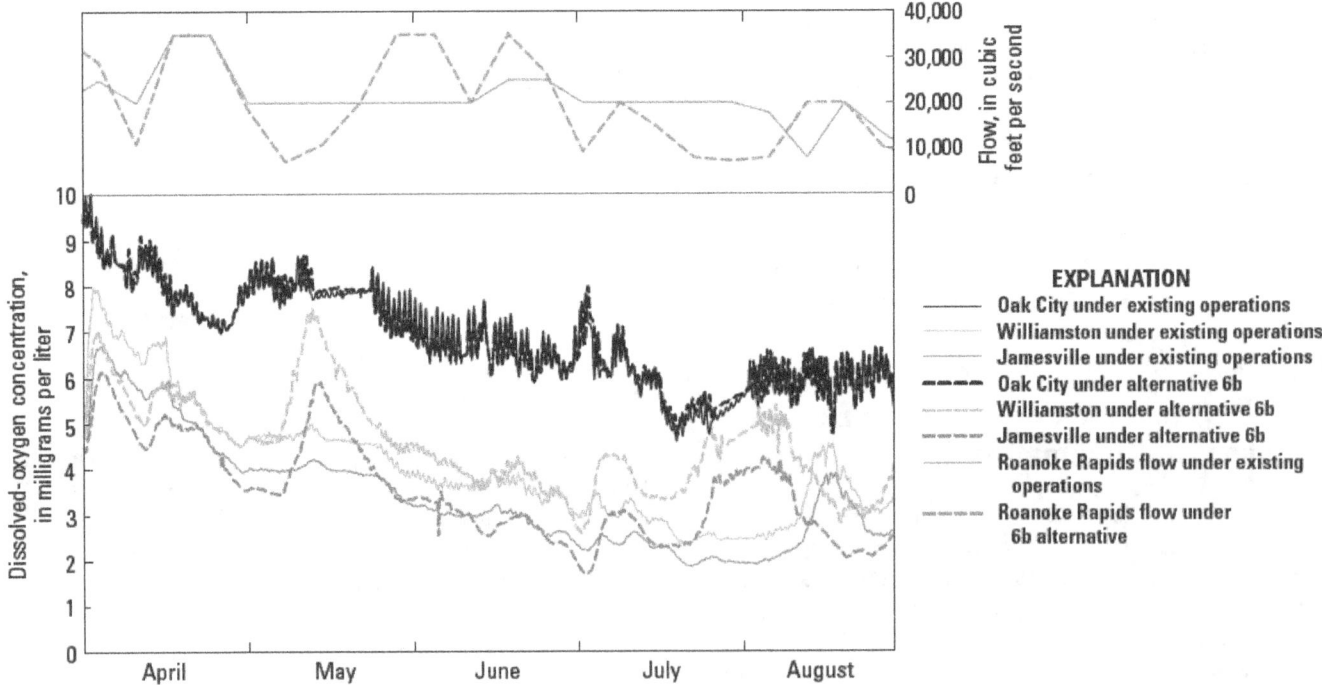

Figure 24.   Simulated dissolved-oxygen concentrations at Roanoke River near Oak City, Williamston, and Jamesville, North Carolina, under existing operations and under alternative 6b.

The greatest effect of alternative 6b on dissolved oxygen relative to existing conditions was predicted at Williamston, which is substantially affected by drainage of upstream flood plains. Mean dissolved-oxygen concentrations at Williamston increased by 10 percent from existing conditions. Under existing operations, dissolved-oxygen concentrations below 5 mg/L occurred 84 percent of the time during the April–August simulation period compared to 72 percent of the time under alternative 6b (fig. 25A).

Alternative 6b not only has the effect of reducing dissolved-oxygen concentrations at sites affected by flood-plain drainage, but also lowers in-stream temperatures (fig. 25B). Maximum temperatures for Williamston, for example, decreased by 3 percent, and minimum temperatures decreased by 6 percent under alternative 6b. These decreases affected the dissolved-oxygen percent saturation (fig. 26). At Williamston, the dissolved-oxygen percent saturation value was about 5–10 percent higher for a given frequency of occurrence throughout the range of values. For example, the median percent saturation for current conditions was 45, whereas under alternative 6b, the median percent saturation was 52.

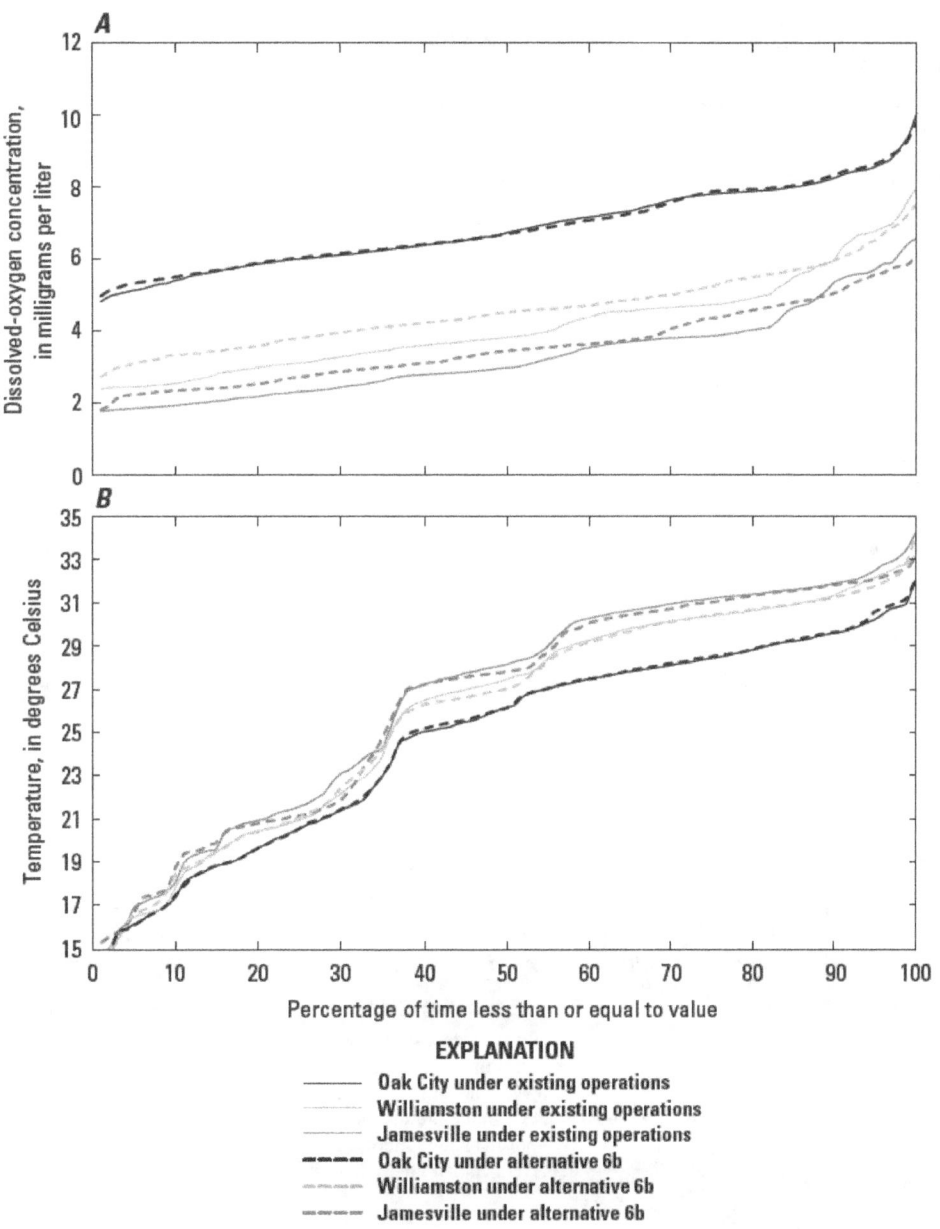

EXPLANATION

——— Oak City under existing operations
········· Williamston under existing operations
——— Jamesville under existing operations
- - - - Oak City under alternative 6b
- - - - - Williamston under alternative 6b
- - - - - Jamesville under alternative 6b

Figure 25. Frequency of occurrence of (A) simulated dissolved-oxygen concentrations and (B) simulated temperature, at Roanoke River near Oak City, Williamston, and Jamesville, North Carolina, under existing operations and under alternative 6b.

Table 9.   Summary statistics of simulated dissolved-oxygen concentrations and exceedances of reference dissolved-oxygen levels at three locations on the Roanoke River for the alternative 6b scenario.

[mg/L, milligrams per liter. Comparisons were done using instantaneous hourly values and using the Federal Energy Regulatory Commission (FERC) minimum flows scenario as a baseline]

	Maximum	Minimum	Mean	Median	Percentage of time dissolved-oxygen concentration is less than	
					5 mg/L[1]	4 mg/L[2]
Oak City						
FERC minimum flows	6.6	5.5	6.0	6.0	0	7
Peaking operations	8.3	4.6	6.0	6.0	7	0
Williamston						
FERC minimum flows	5.7	4.4	4.9	4.8	69	0
Peaking operations	7.1	3.9	5.2	5.1	43	0
Jamesville						
FERC minimum flows	6.3	3.6	4.5	4.2	76	29
Peaking operations	6.7	2.7	4.5	4.5	72	30

[1] North Carolina water-quality daily standard for dissolved oxygen

[2] North Carolina water-quality instantaneous standard for dissolved oxygen

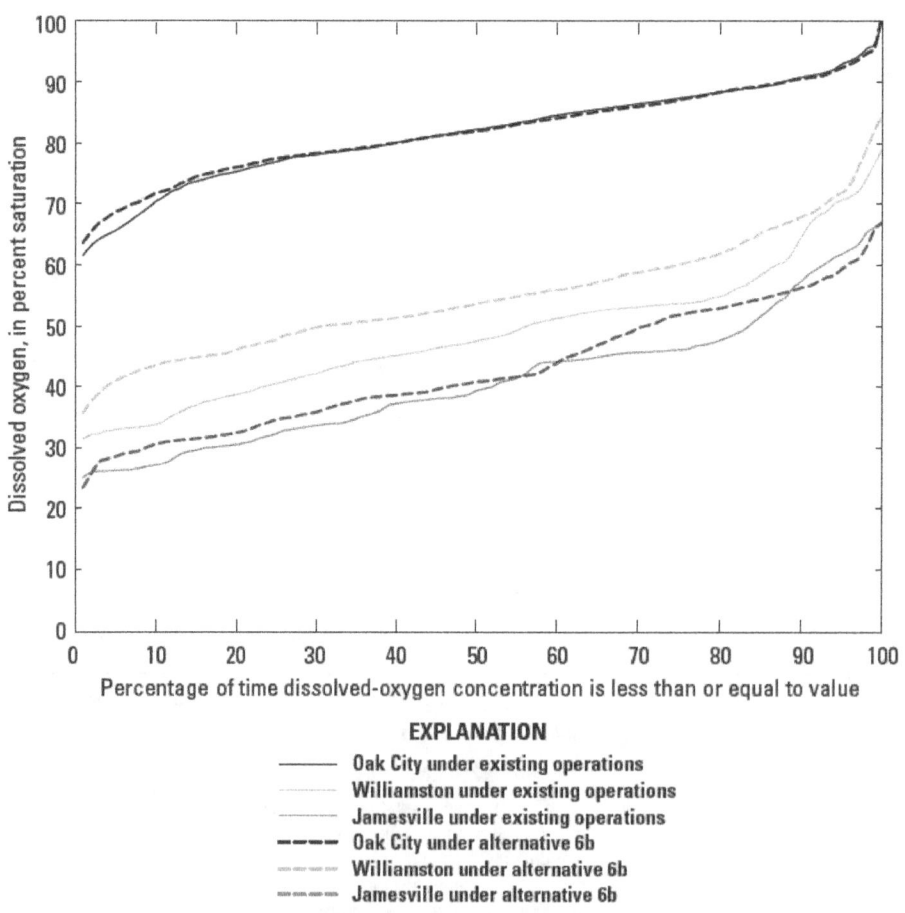

EXPLANATION

————— Oak City under existing operations
··········· Williamston under existing operations
------------ Jamesville under existing operations
— — — — Oak City under alternative 6b
~ ~ ~ ~ Williamston under alternative 6b
– – – – Jamesville under alternative 6b

Figure 26.   Frequency of occurrence of simulated dissolved-oxygen percent saturation at Roanoke River near Oak City, Williamston, and Jamesville, North Carolina, under existing operations and under alternative 6b.

Because of less frequent flooding, the effect of alternative 6b at the Roanoke River near Oak City is small. Near Williamston, however, alternative 6b had a substantial effect on the water quality of both the river and the neighboring flood plain. Operations under alternative 6b have the effect of making inundation of the flood plain more dynamic than under existing operations (figs. 27–29). The flood plain drains rapidly under alternative 6b, which reduces standing time in the flood plain that leads to anoxic conditions. Higher water velocity in the flood plain (fig. 28), which is possible under alternative 6b relative to existing conditions, not only facilitates more rapid drying, it also increases reaeration, and this mechanism leads to improved dissolved-oxygen levels both in the flood plain and in the main stem. The increased interaction with the flood plain results in water being recharged to the main stem that has had less time in the flood plain to become anoxic, thereby improving the dissolved-oxygen concentrations of the lower Roanoke River.

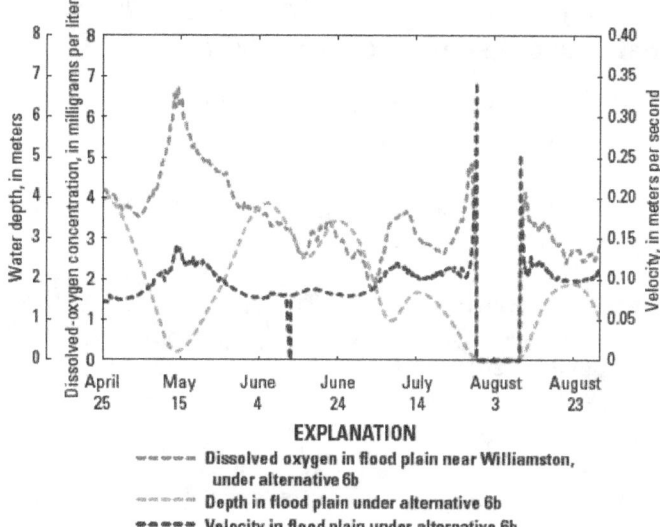

Figure 28. Simulated dissolved-oxygen concentrations, water levels, and velocity at a flood-plain location near Williamston, North Carolina, under alternative 6b.

At the Jamesville location the interaction between the flood plain and the main stem increases, and distinctions are more difficult to make (fig. 29). As a result, the water quality in the main stem aproximates the quality in the flood plain. Also water levels are less dynamic at this location, the differences between reservoir operations are less noticeable, and hence the effect of alternative 6b on water quality in the river is less noticeable.

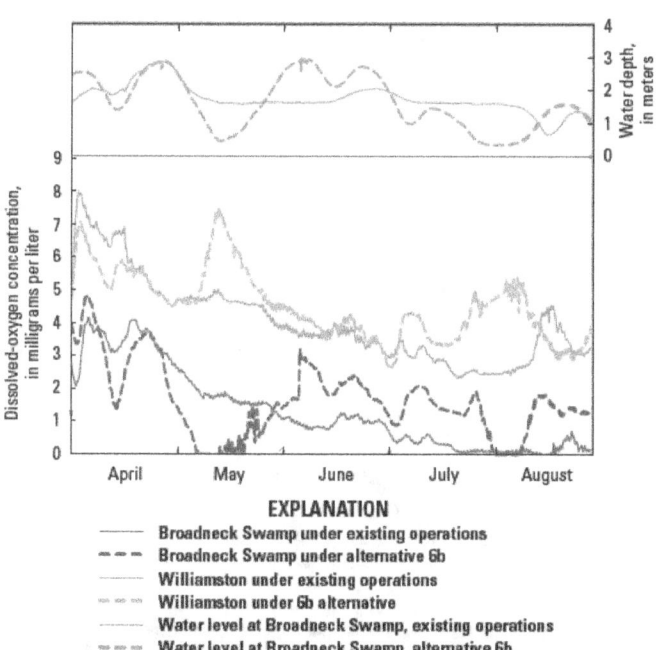

Figure 27. Simulated dissolved-oxygen concentrations and water levels at a swamp location near Broadneck for existing operations and alternative 6b, and dissolved-oxygen concentration at Roanoke River near Williamston, North Carolina.

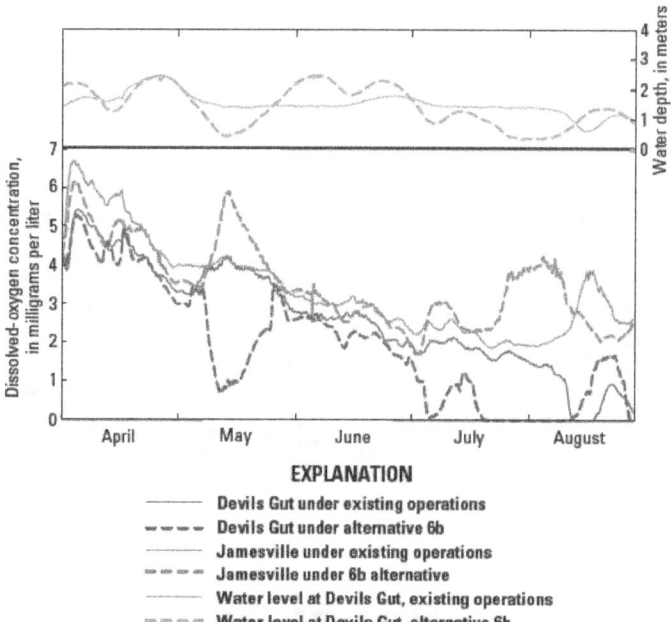

Figure 29. Simulated dissolved-oxygen concentrations and water levels at a swamp location near Devils Gut for existing operations and alternative 6b, and dissolved-oxygen concentration at Roanoke River near Jamesville, North Carolina.

Evaluation of a Dynamic Stepping-Down Release Schedule (Betterment Plan)

In the lower Roanoke River, fish kills have occurred downstream from Oak City, N.C., where drainage of side channels and back swamps that contain lower concentrations of dissolved oxygen have entered into the main channel. A plan to make more gradual flow transitions between flood releases at Roanoke Rapids has been implemented since 2003. This plan is referred to by stakeholders as the betterment plan (U.S. Army Corps of Engineers, 2010) and consists of reducing the flow releases in a step-wise manner. The betterment plan would reduce release flows below 20,000 ft³/s (566 m³/s) in three 5,000-ft³/s (142-m³/s) steps (fig. 30), holding each of those incremental reduction levels for several days. A modified

version of the betterment plan also involves a step-down procedure that includes two steps instead of three. The first step would be a decrease in discharge from 20,000 ft³/s to 12,000 ft³/s (566 to 340 m³/s), a level sustained for 7 days. The second step would lower the discharge to 5,000 ft³/s (142 m³/s) for another 7 days before reaching the minimum flow release levels.

Over the entire summer season the effect of the stepping down procedure is negligible (fig. 31). If, however, the analysis focuses on the weeks during the betterment plan implementation and several days afterward as the effects propagate downstream, then some differences in dissolved-oxygen concentrations are notable (fig. 32). Summary statistics for dissolved-oxygen concentrations at model assessment locations on the Roanoke River are presented in table 10.

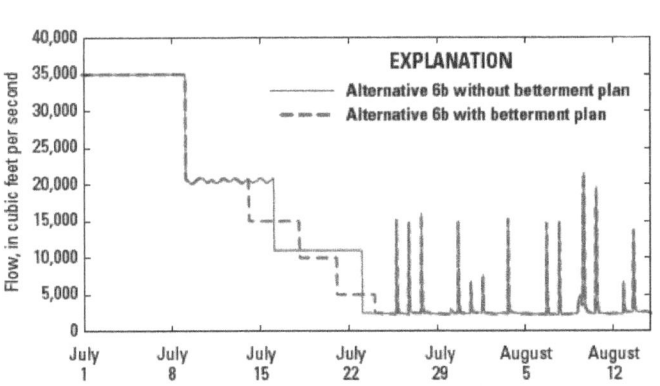

Figure 30. Discharge at Roanoke Rapids under operations with and without a betterment plan for a simulated summer period (July–August). [Data provided by the U.S. Army Corps of Engineers, written commun., 2011.]

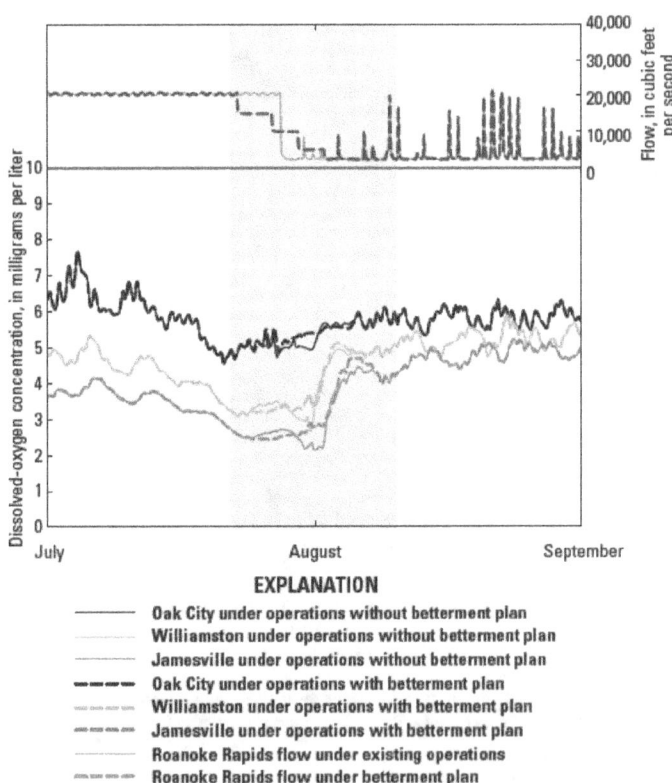

EXPLANATION
——— Oak City under operations without betterment plan
——— Williamston under operations without betterment plan
——— Jamesville under operations without betterment plan
– – – Oak City under operations with betterment plan
– – – Williamston under operations with betterment plan
– – – Jamesville under operations with betterment plan
········ Roanoke Rapids flow under existing operations
– – – Roanoke Rapids flow under betterment plan

Figure 31. Simulated dissolved-oxygen concentrations at Roanoke River near Oak City, Williamston, and Jamesville, North Carolina, under existing operations with and without a betterment plan. The simulated period showing implementation and effects on dissolved oxygen is highlighted in gray.

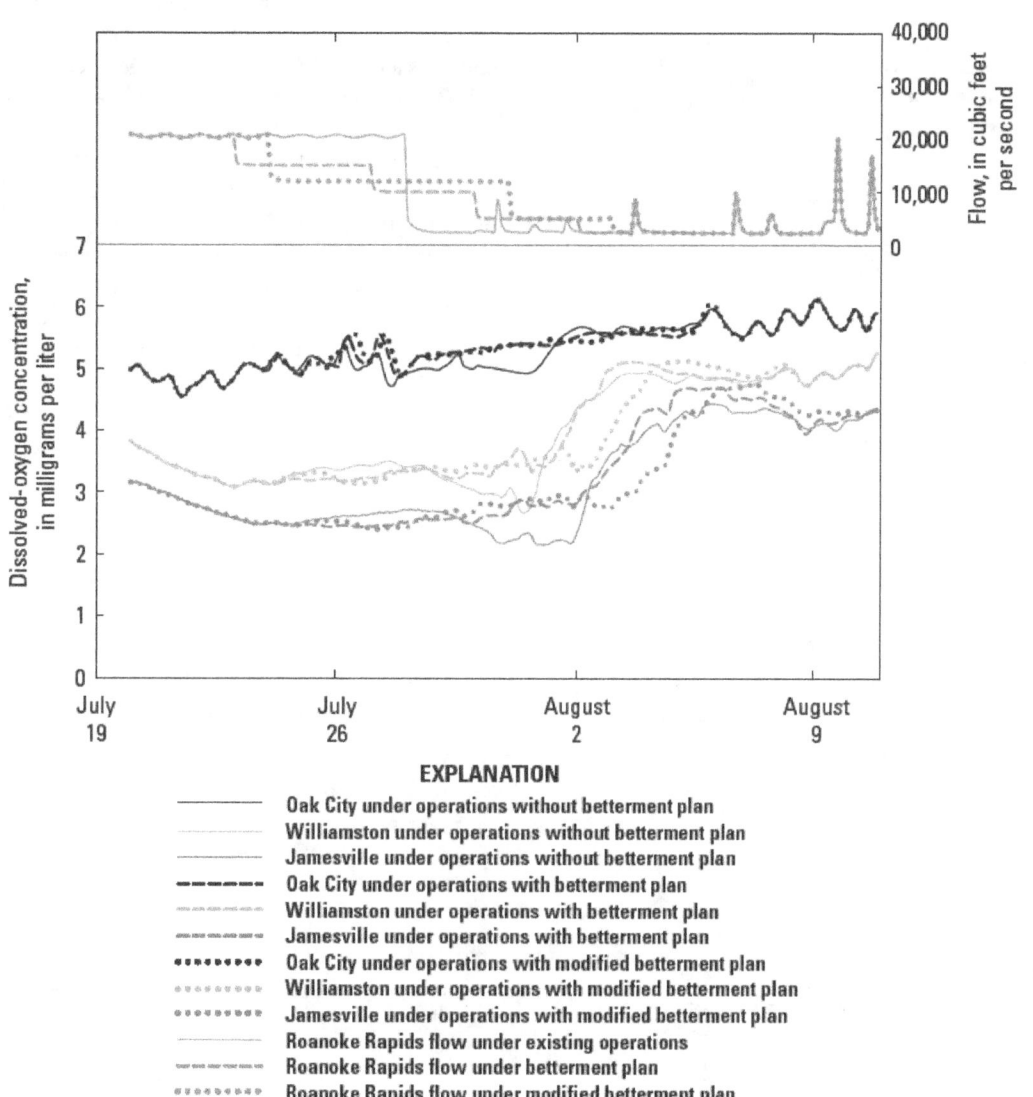

Figure 32. Simulated dissolved-oxygen concentrations at Roanoke River near Oak City, Williamston, and Jamesville, North Carolina, under existing operations with and without a betterment plan, and with a modified betterment plan for weeks of implementation and effect.

Table 10. Summary statistics of simulated dissolved-oxygen concentrations and exceedances of reference dissolved-oxygen levels at three locations on the Roanoke River for the betterment plan scenario.

[mg/L, milligrams per liter. Comparisons were done using instantaneous hourly values and using the Federal Energy Regulatory Commission (FERC) minimum flows scenario as a baseline]

	Maxi-mum	Mini-mum	Mean	Median	Percentage of time dissolved-oxygen concentration is less than	
					5 mg/L[1]	4 mg/L[2]
Oak City						
Operations without betterment plan	6.1	4.5	5.3	5.2	0	19
Operations with betterment plan	6.1	4.5	5.4	5.4	0	0
Operations with modified betterment plan	6.1	4.5	5.4	5.4	0	0
Williamston						
Operations without betterment plan	5.3	2.7	3.9	3.5	94	57
Operations with betterment plan	5.3	3.1	4	3.5	87	58
Operations with modified betterment plan	5.3	3.1	3.9	3.5	83	64
Jamesville						
Operations without betterment plan	4.4	2.2	3.2	2.7	100	69
Operations with betterment plan	4.7	2.4	3.3	2.8	100	67
Operations with modified betterment plan	4.7	2.4	3.2	2.9	100	72

[1] North Carolina water-quality daily standard for dissolved oxygen

[2] North Carolina water-quality instantaneous standard for dissolved oxygen

Maximum and mean concentrations of dissolved oxygen remain unchanged under both versions of the betterment plan. The greatest effect on dissolved-oxygen concentration was predicted at the Williamston assessment location where minimum concentrations increased by 15 percent under both versions of the betterment plan. The number of instances where dissolved-oxygen levels were below 5 mg/L was reduced, but for a greater number of instances dissolved oxygen was less than 4 mg/L. The model predicts a modest effect on dissolved-oxygen concentrations by both versions of the betterment plan; however, the effects are most pronounced at the Williamston assessment location where the interaction with the flood plain is most notable.

The overall effect of the betterment plan, both in the current implementation and proposed modification, is to allow for a more gradual drawdown of flood-plain water levels. Without a betterment plan, existing operations can dry out the flood plain fairly quickly (4 days for the flood-plain location near Williamston, fig. 33). With a step-wise flow reduction at Roanoke Rapids, the period for the flood plain to dry out is increased. This change has several effects on water quality. First, a gradual drawdown allows for reaeration to compensate for the oxygen demand in the flood plain. As a result, the dissolved-oxygen decline (observed on July 31) that results from flood-plain drainage is minimized by a stepping-down procedure.

The second effect of the betterment plan on water quality is to lower maximum water temperatures (fig. 34). As the flood-plain depth decreases, water temperatures increase in the flood plain along with in-stream temperatures. With the gradual drawdown, temperatures in the main channel do not escalate as quickly or as high as without the betterment plan. Thus, the effects of the betterment plan beyond its period of implementation are propagated, and the effects are particularly notable in the case for the modified version where the effects are noted up to 6 days after flows at Roanoke Rapids reach minimum levels.

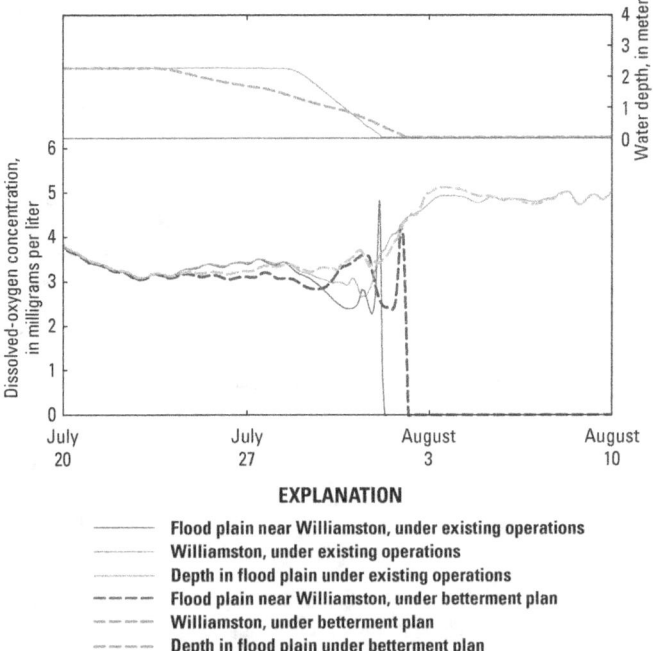

EXPLANATION

——— Flood plain near Williamston, under existing operations
——— Williamston, under existing operations
——— Depth in flood plain under existing operations
– – – Flood plain near Williamston, under betterment plan
– – – Williamston, under betterment plan
– – – Depth in flood plain under betterment plan

Figure 33. Simulated dissolved-oxygen concentrations and water depth at Roanoke River and flood plain near Williamston, North Carolina (Broadneck Swamp), under existing operations and the betterment plan for weeks of implementation and effect (July 20–August 10).

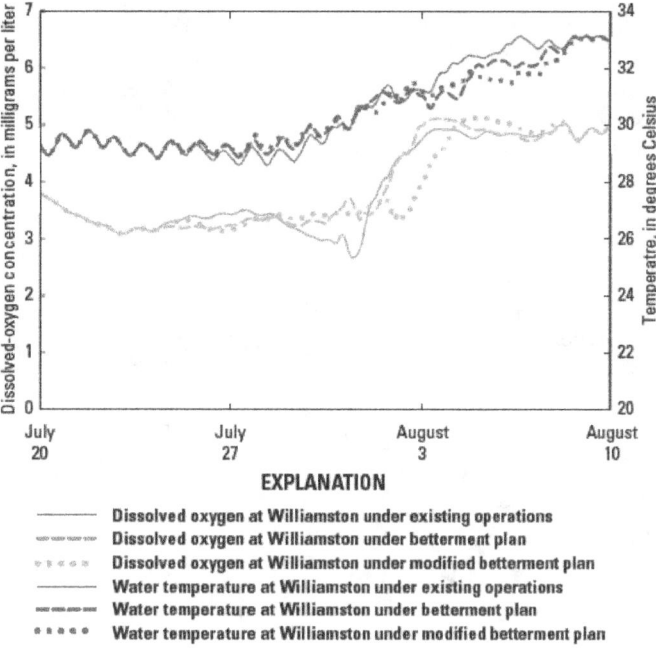

EXPLANATION

—————— Dissolved oxygen at Williamston under existing operations
‑‑‑‑‑‑‑ Dissolved oxygen at Williamston under betterment plan
∙ ∙ ∙ ∙ ∙ Dissolved oxygen at Williamston under modified betterment plan
—————— Water temperature at Williamston under existing operations
— — — Water temperature at Williamston under betterment plan
∙ ∙ ∙ ∙ ∙ Water temperature at Williamston under modified betterment plan

Figure 34. Simulated dissolved-oxygen concentrations and temperature at Roanoke near Williamston, North Carolina, under existing operations, the betterment plan, and the modified betterment plan for weeks of implementation and effect (July 20–August 10).

The alternative 6b flow release schedule also has the goal of minimizing flood-plain drainage effects. The model was used to evaluate how these effects would change if alternative 6b was implemented with a more dynamic stepping-down schedule, specifically the betterment plan. The resulting upstream boundary conditions for this scenario are depicted in figure 35.

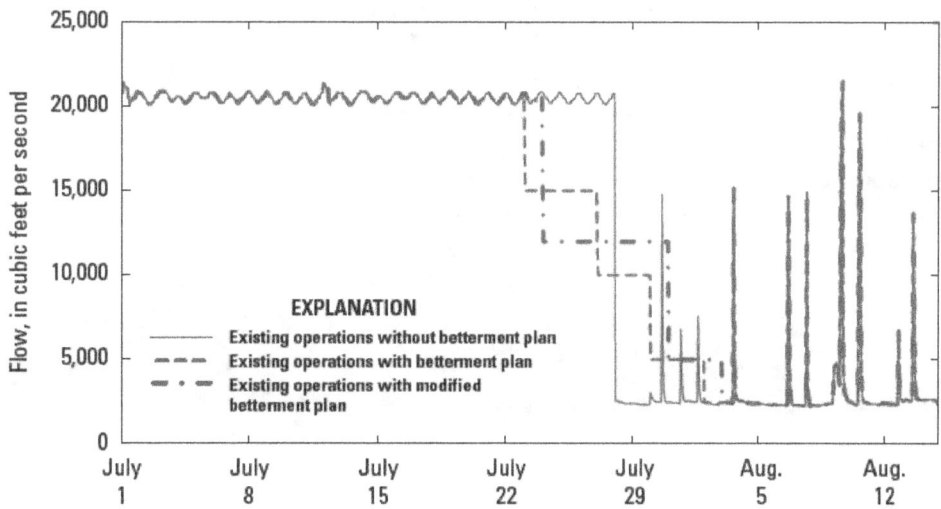

EXPLANATION

—————— Existing operations without betterment plan
‑ ‑ ‑ ‑ Existing operations with betterment plan
— ∙ — Existing operations with modified betterment plan

Figure 35. Discharge at Roanoke Rapids under operations with and without a betterment plan for a simulated summer period (July–August). [Data provided by the U.S. Army Corps of Engineers, written commun., 2011.]

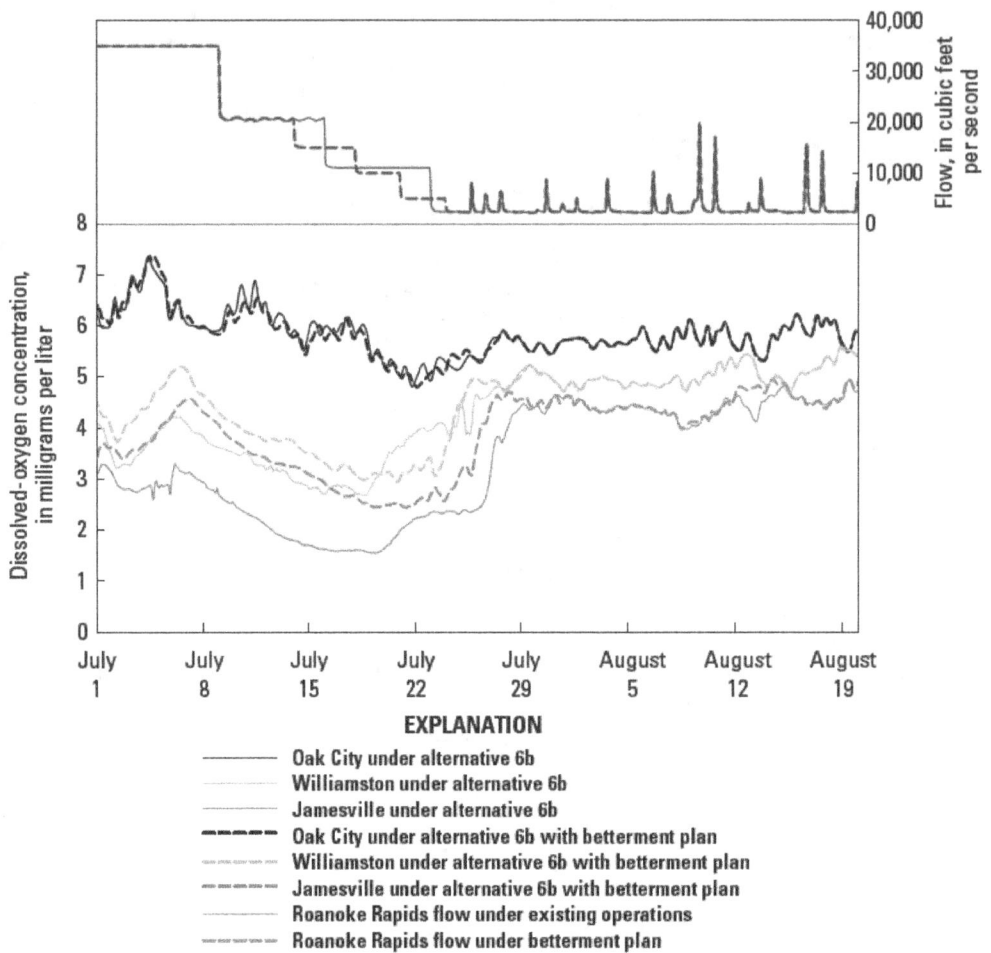

Figure 36. Simulated dissolved-oxygen concentrations at Roanoke River near Oak City, Williamston, and Jamesville, North Carolina, under alternative 6B with and without a betterment plan, for weeks of implementation and effect (July 1–August 20).

The model predicts substantial improvements to in-stream dissolved-oxygen concentrations for the portions of the Roanoke River that are dominated by flood-plain drainage, as represented by the Williamston and Jamesville assessment locations (fig. 36; table 11). Minimum concentrations are increased by 10 and 60 percent at Williamston and Jamesville, respectively, and as a consequence, exceedances of the instantaneous 4 mg/L standard are reduced.

Table 11. Summary statistics of simulated dissolved-oxygen concentrations and exceedances of reference dissolved-oxygen levels at three locations on the Roanoke River for the betterment plan and alternative 6b scenario.

[mg/L, milligrams per liter. Comparisons were done using instantaneous hourly values and using the Federal Energy Regulatory Commission (FERC) minimum flows scenario as a baseline]

	Maximum	Minimum	Mean	Median	Percentage of time dissolved-oxygen concentration is less than	
					5 mg/L[1]	4 mg/L[2]
Oak City						
Alternative 6b	7.4	4.8	5.8	5.8	0	2
Alternative 6b with betterment plan	7.4	4.2	5.8	5.8	0	0
Williamston						
Alternative 6b	5.6	2.7	4.3	4.7	76	40
Alternative 6b with betterment plan	5.7	3.0	4.5	4.8	74	28
Jamesville						
Alternative 6b	5.0	1.5	3.4	3.3	99	53
Alternative 6b with betterment plan	5.1	2.4	3.9	4.3	97	41

[1] North Carolina water-quality daily standard for dissolved oxygen

[2] North Carolina water-quality instantaneous standard for dissolved oxygen

Summary

The Environmental Fluid Dynamics Code (EFDC) and Water Quality Analysis Simulation Program (WASP) models were applied and calibrated to the lower Roanoke River to evaluate several management scenarios, including flood control and hydropower operations. Calibrations were performed for water level, streamflow, and dissolved oxygen for warm-weather months in 2006. The three-dimensional, unsteady numerical model EFDC was used to simulate hydrodynamics and was applied in a depth-averaged mode from Roanoke Rapids, N.C., to Jamesville, N.C. Both longitudinal (upstream to downstream) and lateral (across the channel and into the flood plain) gradients in water depth, velocity, and flow were successfully simulated, and vertical gradients were averaged. The WASP model was configured to use hydrodynamic data generated by EFDC to simulate dissolved-oxygen concentrations. The WASP eutrophication submodel was configured for a simple Streeter-Phelps analysis to simulate dissolved-oxygen flux across the flood plain and along the Roanoke River.

Calibration of the hydrodynamics and dissolved-oxygen concentrations emphasized the effect that flood-plain drainage has on in-stream dissolved-oxygen levels. The hydrodynamic calibration was considered successful because in-stream streamflow and water levels were well predicted, and water-level fluctuations at a location on the flood plain were reasonably represented. The WASP model can also be considered satisfactory and was shown to simulate the effect of flood-plain drainage on in-stream dissolved-oxygen concentrations.

The scenarios that most influenced water quality were those that changed the dynamics of flood-plain drainage as well, including revised release schedules, such as alternative 6b. If no changes are made to the exisisting operations for Kerr Dam, most water-quality issues will remain largely unchanged.

References

Ambrose, R.B., Wool, T.A., Martin, J.L., 1993, The Water Quality Analysis Simulation Program, WASP5 version 5.10, Part A—Model documentation: Athens, Ga., U.S. Environmental Protection Agency, Office of Research and Development, Environmental Research Laboratory.

Bales, J.D., Strickland, A.G., and Garrett, R.G., 1993, An interim report on flows in the lower Roanoke River, and water quality and hydrodynamics of Albemarle Sound, North Carolina, October 1989–April 1991: U.S. Geological Survey Open-File Report 92–123, 133 p.

Bales, J.D., Tomlinson, S.A., and Tillis, Gina, 2006, Flow and salt transport in the Suwannee River estuary, Florida, 1999–2000, Analysis of data and three-dimensional simulations: U.S. Geological Survey Professional Paper 1656–B, 66 p.

Bales, J.D., and Walters, D.A., 2003, Relations among flood-plain water levels, instream dissolved-oxygen conditions, and streamflow in the lower Roanoke River, North Carolina, 1997–2001: U.S. Geological Survey Water-Resources Investigations Report 03–4295, 81 p.

Hamrick, J.M., 1992, A three-dimensional environmental fluid dynamics computer code—Theoretical and computational aspects: Gloucester, Va., The College of William and Mary, Virginia Institute of Marine Science, Special Report 317, 63 p.

Ji, Z.-G., Morton, M.R., and Hamrick, J.M., 2001, Wetting and drying simulation of estuarine processes: Estuarine, Coastal and Shelf Science, v. 53, p. 683–700.

Jin, K.-R., Hamrick, J.H., and Tisdale, T., 2000, Application of a three-dimensional hydrodynamic model for Lake Okeechobee: Journal of Hydraulic Engineering, v. 126, no. 10, p. 758–771.

Jin, K.-R., Ji, Z.-G., and Hamrick, J.H., 2002, Modeling winter circulation in Lake Okeechobee, Florida: Journal of Waterway, Port, Coastal, and Ocean Engineering, v. 128, no. 3, p. 114–125.

Lebo, M.E., 1998, Roanoke River studies—1997: New Bern, N.C., Weyerhaeuser, Southern Environmental Field Station, project no. 722–9812, 7 p.

Mellor, G.L., and Yamada, Tetsuji, 1982, Development of a turbulence closure model for geophysical fluid problems: Reviews of Geophysics, v. 20, p. 851–875.

Moriasi, D.N., Arnold, J.G., Van Liew, M.W., Binger, R.L., Harmel, R.D., and Veith., T.L., 2007, Model evaluation guidelines for systematic quantification of accuracy in watershed simulations: Transactions of the American Society of Agricultural and Biological Engineers, v. 50, no. 3, p. 885–900.

Moustafa, M.Z., and Hamrick, J.M., 1994, Modeling circulation and salinity transport in the Indian River Lagoon *in* Spaulding, M.L., Bedford, K., Blumberg, A., Cheng, R., and Swanson, C., eds., Estuarine and Coastal Modeling, Proceedings of the 3d International Conference: New York, American Society of Civil Engineers, p. 381–395.

Nash, J.E., and Sutcliffe, J.V., 1970, River flow forecasting through conceptual models, part I—A discussion of principles: Journal of Hydrology, v. 10, no. 3, p. 282–290.

North Carolina Division of Emergency Management, 2002, North Carolina floodplain mapping—Pasquotank Basin, 20 ft digital elevation model: Cary, N.C., Floodplain Mapping Program.

Suscy, P., and Morris, F.W., 1998, Proposed discharge limit for Turkey Creek, Brevard County for maintaining a desirable salinity regime in Indian River Lagoon: Palatka, Fla., St. Johns River Water Management District, Technical Memorandum No. 26, 174 p.

Todd, M.J., Vellidis, George, Lowrance, R.R., and Pringle, C.M., 2009, High sediment oxygen demand within an instream swamp in southern Georgia—Implications for low dissolved oxygen levels in coastal blackwater streams: Journal of the American Water Resources Association, v. 45, no. 6, p. 1493–1507 (also available at *http://dx.doi.org/10.1111/j.1752-1688.2009.00380.x*).

Townsend, P.A., 2000, A quantitative fuzzy approach to assess mapped vegetation classifications for ecological applications: Remote Sensing of Environment, v. 72, p. 253–267.

U.S. Army Corps of Engineers, 2010, John H. Kerr Dam and Reservoir Feasibility Study: Wilmington, N.C., Wilmington District, Feasibility Scoping Meeting Report, p. 31.

U.S. Geological Survey, 2007a, Facing tomorrow's challenges—U.S. Geological Survey science in the decade 2007–2017: U.S. Geological Survey Circular 1309, 70 p. (Accessed April 12, 2012, at *http://pubs.er.usgs.gov/ usgspubs/cir/cir1309.*)

U.S. Geological Survey, 2007b, The Cooperative Water Program—Program priorities for 2007: Water Resources Discipline Informational Memorandum No. 2007.01. (Accessed March 10, 2008, at *http://water.usgs.gov/coop/ priorities.html.*)

Wehmeyer, L.L., and Wagner, C.R., 2011, Relation between flows and dissolved oxygen in the Roanoke River between Roanoke Rapids Dam and Jamesville, North Carolina, 2005–2009: U.S. Geological Survey Scientific Investigations Report 2011–5040, 29 p.

Wool, T.A., Davie, S.R., and Rodriguez, H.N., 2003, Development of three-dimensional hydrodynamic and water quality models to support total maximum daily load decision processes for the Neuse River estuary, North Carolina: Journal of Water Resources Planning and Management, v. 129, no. 4, p. 295–306.